OUTLINES

OF

MUSIC HALL LECTURES,

EMBRACING FIVE ADDRESSES ON

FACTORY REFORM

IN

THE LARGEST TRADE OF THE UNITED STATES.

DELIVERED AT MUSIC HALL, LYNN, MASS., 1871.

BY

JOSEPH COOK.

The Nineteenth is the Century of the working men. — GLADSTONE.

L'aristocratie manufacturière de nos jours, après avoir appauvri et abruti les hommes dont elle se sert, les livre en temps de crise à la charité publique, pour les nourir. — DE TOCQUEVILLE.

Εἰς τ' ἀκάθαρτα λοιμὸς ἐσοικίζεται. — MENANDER.

BOSTON.
W. H. HALLIDAY AND COMPANY.
1871.

THE MUSIC HALL MEETINGS.

MR. COOK, we learn, is to be at Cambridge and Andover for a few weeks, before leaving for Europe. The following letter to him, which has been numerously signed, explains itself. On account of the appearance of the request contained therein, Mr. Cook has withheld from local publication the usual and promised report of his remarks, in order that they may appear for the first time in pamphlet form:

LYNN, May 11, 1871.

REV. JOSEPH COOK:—*Dear Sir:* Now that the Sabbath evening meetings held by the First Congregational Church in Music Hall, and attended by extraordinarily large audiences, are ended; and because of the importance of the questions there discussed, and the relations they bear to the future growth, interests, and general welfare of Lynn; and also desiring to have preserved in a convenient and permanent form the various phases of excitement and discussion, during the winter, upon matters of great importance to the general welfare; and as a fitting and profitable close to your labors for the public good in this city; the undersigned most respectfully request that, if convenient, you will publish in pamphlet form the outlines of all your Sunday Evening Lectures in Music Hall, through the months of January, February, March, and April, and also all of the cards, letters, and other papers pertaining to the discussion of these subjects, which have from time to time appeared in the city papers. Hoping, dear sir, that you will, if possible, comply with the terms of this request, and thus crown your efforts for good in this city, and with best wishes for your continued health and prosperity, we remain,

Very Respectfully, Yours, etc.

The Transcript, May 13.

Except the first and last of the five Addresses on Factory Reform in the Largest Trade of the United States, this publication is a republication. The reader will find evidence in the pamphlet that the local hurricane caused by the delivery of these addresses was overwhelmingly in favor of Factory Reform. The local public sentiment here exhibited in one of the most important manufacturing centres of New England is a great credit to the city. It is, also, an important argument in relation to a question of high public interest; and it is as such that, in a pamphlet intended chiefly for local circulation, the record of it has been preserved. In compliance with the foregoing request, the compilation has been made complete, and is in no sense partisan. Except a few slight and empty newspaper paragraphs, the omission of which no one could think of any importance, nothing whatever that could illustrate public sentiment has been omitted. The excellent reports of the *Transcript* are omitted, except in two or three instances, as they contain only matter parallel with those of the *Reporter*, to which the courtesy of a reprint is here extended. A correct statement as to the authorship of these reports will be found on p. 128. In the first and fifth Addresses on Factory Reform, originally delivered extemporaneously, a few statistical and historical illustrations have been added; but these occur at points not criticised, the points criticised remaining in substance, and, so far as the words could possibly be recalled, in language, unaltered.

ANDOVER, MASS.

CONTENTS.

Request for Publication, ... iii
Statement in Reply to Request, ... iv

LECTURE I.

Similarity of Feeling with God, a Condition of Salvation, . 1

LECTURE II.

The Moral Perils of the Present Factory System of Lynn, . 3
First Address on Factory Reform in the Largest Trade of the United States, ... 4

LECTURE III.

False and True Faith and Repentance, ... 37
Second Address on Factory Reform, ... 37
Remarks by Mr. S. M. Bubier at Music Hall, . . . 43
Card from the First Congregational Church and Society, . 44
Card from Eight Female and Six Male Employees, from Mr. Henry Downing, and from Messrs. Frazier and Bubier, ... 44
Other Reports by the Local Press, ... 45
Card from Mr. S. M. Bubier, ... 49

LECTURE IV.

Christ's Gift of the Fulness of Joy; or, the First Principles of a Scientific Christianity, ... 50
Third Address on Factory Reform, ... 50
Article by the Rev. Patrick Strain, ... 59
Card from Messrs. Berry and Beede, ... 61
Reply by "A Hearer," 61

LECTURE V.

Treason toward God; or, Christ's Human Nature the Ideal of Excellence for Young Men, ... 63

LECTURE VI.

Secret Sins as an Obstacle to Conversion, ... 64
Remarks against Applause at Music Hall, ... 64
Address by Mr. George W. Keene, ... 65
Remarks in Reply to Mr. Keene by Messrs. Moore, Bancroft, Roberts, Butler, and Carruthers, ... 73
Article on "The License of the Pulpit," 76
Reply by "Justice," 78

LECTURE VII.

Poor Clothes, Rear Pews, and Hard Work as Excuses for not attending Church; or, the Value to the People of Sunday Kept Holy, ... 80
Music Hall Crowd, ... 83

LECTURE VIII.

The Necessity of the Atonement, an Inference from the Nature of Conscience and from the Unchangeableness of the Past, . 84
Fourth Address on Factory Reform, ... 84

LECTURE IX.

Christ in the Family and in all the Relations of the Social to Religious Life; or, Cheap Homes and Boarding-Houses in Lynn, ... 97

LECTURE X.

Steps toward the New Birth; or, Recent Science on the Signs of the Vices, and on the best Methods of Escape from Bad Habits, ... 103
Petition with Signatures of Two Hundred Working Men, 108
Answer to the Request, ... 109

LECTURE XI.

Swindling Public Amusements and the Street School; or, Christianity and Science at one as to the Temptations of Cities, ... 110
Remarks on the Need of a City Missionary in Lynn, . 110

LECTURE XII.

Club-Rooms in Lynn, Gambling-Rooms, and Dwelling-house Sabbath Drinking Parties; or, the Duty of making Christian Public Sentiment in Cities a Police behind the Police, 120
Second Remarks on the Need of a City Missionary in Lynn, ... 116
Resolutions of the Committee having Charge of the Music Hall Services, ... 118

LECTURE XIII.

Christ our Prophet, Priest, and King; or, Christ not a Saviour if worshipped in only a Fragment of His Character, . 126
Expenses of the Music Hall Meetings, ... 126

Ocean Sunrise from High Rock, ... 127
Subjects and Dates of Music Hall Lectures, . . . 127
Authorship of Reports of the Meetings, . . . 128
Fifth Address on Factory Reform, ... 128
The Thirteenth of the Music Hall Services, . . . 141
Testimonial from Members of the First Congregational Church and Society, ... 143
Card from Mr. S. M. Bubier, ... 143
A Question of Privilege, ... 144
Reply by the Editor of the *Reporter*, ... 146
Hearing with the Elbows, ... 147
Letter of Sept. 7, 1871, from Working Men, ... 147
Extracts on Children in Factories and the Factory System at Lynn, from the Second Report of the Massachusetts Bureau of Statistics of Labor, ... 147

LECTURE I.

SUNDAY EVENING SERVICES AT MUSIC HALL.

THE religious services which took place in Music Hall last Sunday evening, under the auspices of the First Congregational Society of this city and in accordance with the announcement in our last issue, were in every sense a perfect success. The meetings were inaugurated as an experiment; and, from the encouragement shown on the above occasion, it is to be presumed that the series will be continued.

Notwithstanding the very inauspicious state of the weather, the hall was filled at an early hour; many being at the door some time previous to the opening of the hall, waiting for an entrance. The services were commenced with voluntary singing by a chorus of voices from the Lynn Choral Union and other vocalists of the city, numbering about sixty. It was expected that one hundred and fifty singers would have been present; but the weather prevented many from coming who would otherwise have appeared. After the reading of selections from the Scripture, prayer, and the singing of "Coronation," and another hymn by the choir, the preacher, Rev. Joseph Cook, announced as his text the words found in John iii. 5: "Jesus answered, Verily, verily, I say unto thee, except a man be born of water and of the Spirit, he cannot enter into the kingdom of God."

The speaker proceeded to show that religion is regarded too much as something pertaining only to Sabbath-day worship, when on the contrary it should be carried into every day life, the business and avocations of every person. The importance of a pure life in accordance with the injunctions of the gospel was shown up in vivid language, and the hearers urged to live in such a manner that they could pass unharmed the breathing holes of perdition which were to be seen on all sides as we walk through the streets of the city. Man in his fallen state is not fit for the kingdom of heaven; and, even could he gain admittance there all vile as he is in a sinful condition, he could not enjoy the pure pleasures of that holy place. This was illustrated by picturing a beautiful woman, pure, holy, and perfect in every way, and then taking a man all polluted with crime and vice, and placing him in the presence of her whose character was spotless, whose every word and act was free from reproach. As such a man would be miserable in so pure a presence, so would the person be in the realms of the blest, unless he had experienced the baptism of the Holy Spirit.

The young men were urged to shun club-rooms as the swimmer would steer clear of the shark; and young ladies were cautioned to avoid with the same care, the company of those who frequent such places. Our evil deeds were pictures which could not be turned face to the wall, and nothing could blot out past transgressions. The subject was continued with great earnestness and elicited the profound attention of his hearers.

After announcing another meeting in the same place next Sabbath evening, when he would preach upon the subject of " The Moral Perils of the Factory System of Lynn," and remarking that he would like to have all present, manufacturers as well as employees, Mr. Cook requested the congregation to join in singing the closing hymn, " Rock of Ages," which was done, and the audience were dismissed with the benediction. A collection was taken up during the services to assist in defraying the expense attending the hiring of the hall. The singing was exceedingly well executed, and added vastly to the interest of the exercises.

We are of opinion that these meetings will be prolific of much good, as many will be brought into the range of religious influences who would otherwise never be present at such assemblies. There will be no postponement of the next service on account of weather. — *Semi-Weekly Reporter*, Jan. 18.

LECTURE II.

SUNDAY EVENING SERVICES AT MUSIC HALL.

THE second meeting for religious services, at Music Hall, in this city, under the auspices of the First Congregational Church, took place last Sunday evening, and was very fully attended, every seat in the hall being filled, while some two or three hundred remained standing, a large number going away without obtaining entrance.

A chorus of about one hundred voices from the Lynn Choral Union, and others of the musical portion of the community, furnished the vocal melody in a most acceptable manner.

After the usual opening exercises, including prayer, reading of the Scriptures, and singing, the speaker, Rev. Joseph Cook, proceeded to consider the subject previously announced: "The Moral Perils of the Present Factory System of Lynn," basing his remarks upon words found in Matthew vi. 13 and James iii. 6. He commenced by saying that he should doubtless be severely criticised in what he said; but the subject was one which he had regarded seriously and with much thought, and he should treat it independently, not with a desire to find fault, but with a view of calling the attention of the thinking portion of the community to what he considered an evil. By this, he meant the practice which was adopted at the present day, of male and female operatives being employed in one room, whereas, in the speaker's judgment, a separation of the sexes was demanded. He would not be understood as saying that Lynn was the seat of all the evil in the State. Old Lynn was just the same as ever in its moral standing. But the transition state of the shoe business from the old to the present factory system, had called in a floating population, which had changed the general character of the place. The present method of doing business necessitated the employment of large numbers in one building, men and women being oftentimes found at work in the same room, which he regarded as detrimental to their moral welfare. The main point urged was a separation of the sexes, or else the most stringent regulations in regard to the characters of those employed. He also alluded to the moral standing of those who had the superintendence of the workshops, insisting upon the importance of having men of irreproachable character to oversee each department. The subject was handled in an open, bold, vigorous manner; and, although many differed from the speaker in his views on the subject, several points made by him were received with marked tokens of appreciation, which culminated in open applause. The audience paid strict attention to the discourse, of which we refrain from giving an extended report, as it is the design to have it put up in cheap pamphlet form for the use of those who are interested in the subject. It will be ready in the course of a few days. — *Semi-Weekly Reporter*, Jan. 25.

EXTRAORDINARY RECENT GROWTH OF LYNN.

THIS city has been greatly prospered in the last twenty years. In 1850 your population was thirteen thousand. Now it is twenty-nine thousand. In 1850 the valuation of your property was four million dollars. Now it is twenty million dollars. For the sixty years from 1795 to 1855, the increase of the value of the productions of your great branch of industry was, every ten years, one hundred per cent. For the ten years from 1855 to 1865, the increase of that value was three hundred per cent.[1] When the soldiers marched, Lynn clothed their feet; and the steel-clad hoof of war, smiting on the Lynn lap-stone, struck forth here abundant sparks of prosperity.

I blame no man for being proud of the recent secular advances of Lynn. The time has been when there was altogether too little local pride here. One of the speakers at the dedication of your city hall, an ex-mayor, said that the local remark used to be that all Lynn was mortgaged to a neighboring city. That day has passed. There is a New Lynn. It is visible enough in the midst of Old Lynn. The city hall costing three hundred and fifty thousand dollars, new school-buildings, new churches, new business blocks, new factories, fifty-five railway trains a day, have made a New Lynn. And Old Lynn, with its poorer architecture, lies in the arms of New Lynn, as the old moon, which I saw in the west as I came to this meeting, lies in the arms of the crescent new moon: a crescent, indeed, as yet, but sure to become a sphere.

I am not unmindful of the attractions which culture and refinement have thrown about your summer resorts. Nahant is the beautiful white arm which Lynn thrusts out into the sea; and it is inspiriting to me to remember that among the rings on the hand of that arm are the summer cottages of

[1] Census of Massachusetts, 1865.

Longfellow and Prescott and Agassiz. There is a cherry-tree on grounds near the side of one of your streets, from which I have plucked a twig with reverence, because around that tree Prescott wore a path by pacing to and fro. There is that which will make the gleam of Massachusetts Bay go with me into whatever land I wander. There is that which will make the roar of the ocean on Long Beach come back to me, it may be even in the ear of death.

I care for Lynn. I recognize its prosperity and its general good order. But, in spite of this, for what I say to-night I expect to be cut into more pieces than you ever cut leather into in your factories.

THE RIGHT ORGANIZATION OF THE LARGEST TRADE OF THE UNITED STATES, OF HARDLY LESS THAN NATIONAL IMPORTANCE.

Wholly aside, however, from the local importance of the extraordinarily delicate and complicated theme which I am to bring before you, there are other and far larger considerations which make its discussion, and its discussion at the present time, of high moment.

Five reasons exist why the right organization of the new factory system of the vast trade of this city, is of altogether more than local, and of hardly less than national importance. The circumstances need only to be named to exhibit their own commanding significance. My personal motives for entering upon the discussion of this evening will undoubtedly be variously construed. I profess that I have no other motives than those which are drawn from the five circumstances I am about to name. I am sure that I need no other and no greater.

1. There are more people engaged in the shoe trade than in any other single branch of manufactures in the United States, not excepting the coal, the iron, the woollen, or the cotton.

The unpoetic shoe trade happens to be the largest single branch of manufactures in the United States. "The manufacture of boots and shoes," says the preliminary report of the Eighth Census, "employs a larger number of operatives than any other single branch of American industry."[1] It is difficult to understand the importance of this fact except by a few statistical contrasts. Take the coal, the iron, the woollen, and the cotton trade. Every one recognizes these as vast and permanent public interests. They fill the land with the noise of hammers and spindles. Whatever affects powerfully the interests of either of these vast branches of industry is a matter of little less than national importance. The right organization of the industrial life of either of these trades would be a problem of commanding weight and interest. But the shoe trade employs a greater number of operatives than either of these other vast trades, taken alone. I find that in 1860 the entire list of woollen manufactures in the United States employed only 48,900 operatives. Of these, 28,780 were males and 20,120 females. The number of operatives employed in the cotton manufactures of the United States in 1860, I find, was only 118,920. Of these, 45,315 were males and 73,605 females. But, by the same census which gives these estimates, the number of hands employed in the United States in the manufacture of boots and shoes in 1860 was 127,427. Of these, 96,287 were males and 31,140 females. Even before the immense increase which the war brought to the shoe trade, the contrast of that branch of manufactures with the cotton, gives predominance to the former, by the difference between 118,920 operatives on the one hand and 127,427 on the other.[2]

In Massachusetts, taken alone, the shoe trade is vastly larger than any other; and this in spite of the fact that Lawrence and Lowell lead the cotton manufactures of the whole nation. This commonwealth in 1860 employed in the manufacture of cotton goods, 34,988 operatives. Of these, 12,635 were males and 22,353 were females. In the manufacture of boots and shoes it employed, in the same year, 69,398 operatives. Of these, 47,353 were males and 22,045 females.[3]

[1] Preliminary Report on the Eighth Census of the United States, p. 68.

[2] Ibid., Tables No. 22, No. 23, and No. 25, pp. 65–69, 180–185.

[3] Ibid., pp. 180, 185.

This largest trade of the United States is rapidly increasing in size. In the journey of growth it moved its tents with vast strides, not only during the war, but before. It is an exceedingly significant fact that, in 1860, 2,554 establishments engaged in this manufacture in the New England States employed a capital only $2,516 less than the whole Union employed in the manufacture in 1850.[1] The entire value of the business of these establishments in 1860 was 82.8 per cent in excess of its value in 1850. New England, New York, Pennsylvania, and New Jersey together produced in 1860, 67.9 per cent more in value than in 1850. In this manufacture Massachusetts increased the value of its products, between 1850 and 1860, at the astonishing rate of 92.6 per cent.[2] I give these facts to exhibit the prosperity of the trade immediately before the war. But it is notorious that during the war, under the double impulse of the demands of the armies and the invention of new machinery for the manufacture, its rate of progress was yet more surprising. It has not shown any want of prosperity since the war. When the products of this trade in 1865 are put into contrast with those of 1870, the figures in Philadelphia are 605,329 compared with 2,111,024; and, in this city, 2,373,203 compared with 6,356,166.

Here, therefore, is a vast public interest; and whoever examines it is not discussing a trivial theme. It is exceedingly important, however, to notice not only that the theme is important, but that its discussion at the present, rather than at a later, period, has an importance second only to that of the theme itself.

THE PRESENT A PERIOD OF TRANSITION TO A WHOLLY NEW FACTORY SYSTEM.

2. On account of the recent invention of new machinery for its processes, this whole vast branch of manufactures throughout the United States is now in a condition of transition from the small shop system to the large factory system; it is adopting industrial arrangements which are likely to be a precedent

[1] Eighth Census, Preliminary Report, p. 68. [2] Ibid., p. 68. Also Table 25.

for an extended and most important future ; and the present, therefore, is the time to strike for the right crystalization of a factory system affecting the industrial and moral interests of a larger number of people than are reached by any other single branch of manufactures in the United States.

RAPIDITY AND EXTENT OF THE CHANGES NECESSITATED BY RECENT MECHANICAL INVENTIONS.

It is a matter of public notoriety that within the last ten years the methods of the shoe manufactures have been revolutionized by the invention of the McKay sewing machine. The invention of the spinning jenny and of the power loom did no more to revolutionize the cotton manufacture; the invention of the steam engine no more to change the methods of inland and maratime conveyance, than the application of the sewing machine to the shoe trade has done to revolutionize the processes of that branch of industry. The change has been as remarkable for rapidity as for extent. It was hastened by the great exigencies of our civil war. The celebrated machine which is likely to be remembered in history side by side with the spinning jenny and the power loom, was invented and patented by Lyman R. Blake of South Abington, in this Commonwealth, as late as the year 1858.[1] When the civil struggle began, it was seen that machinery must do the work of the multitudes of mechanics of the North, who had left their places and were fighting the battles of the war. The original patent was sold to Mr. Gordon McKay and Mr. J. G. Bates of Boston, for ten thousand dollars. It was somewhat improved by them. Not far from the second year of the war, it began to be applied to the shoe manufactures in establishments in this city. It is unnecessary to recite the rest of the history. Invention has followed invention. The supply of the wants of the new system of factories has tasked the skill of the best experts in machinery in New England. The McKay sewing machine, the skiving machine, the pegging machine, the sole-molding machine, the cable-wire machine,

[1] Shoe and Leather Record, Boston, Sept. 26, 1870.

the self-feeding eyelet machine, are but a fraction of the recent inventions not only patented, but in use. Any list of machines correct for to-day is likely to be incorrect, because outgrown, to-morrow. Rapid as the supply of the new machinery has been, the demand for it has exceeded, and yet exceeds, the supply.

Three large results have followed this invention of new machinery. First, the small shop system has been abandoned and the large factory system has been adopted. Secondly, a great subdivison of labor has taken place. Thirdly, the trade is much more subject to lulls, or inactive seasons, than it was before the invention of the new machinery.

Occurring in the largest trade of the United States, these changes are events of a high order of public importance.

The transition from the old system to the new is complete and final. All Eastern Massachusetts is sprinkled thick with the small shoe shops, buildings twelve or twenty feet square, in each of which ten or fifteen men were usually employed on the heavier work of the trade, the females, in their own rooms at home, doing the lighter work. These buildings have been vacated, never to be filled again. For a hundred years they have been almost as characteristic of a large part of the towns of Eastern Massachusetts as the schoolhouses or the churches. The large factories, which are rising to fill their places, are destined to become larger and larger. There is no longer any artisan in this trade who makes a whole shoe. Subdivision of labor is sometimes carried so far that a single article passes through the hands of fifty workmen, each of whom is trained only to make a part. As a rule, the old shoemakers were largely independent in the management of their business, each family attending to its own for itself. But the large factories have introduced an operative class and an employing class. In the old system, work was commonly steady from year's end to year's end; or affected only by the larger fluctuations of general commerce. But now there are two periods in each year in the trade in any large city when hundreds of operatives are dropped from employment. So far apart at so many points are the old system and the new, that it is of little service to reason from the experience of the trade under the former system to the experience it is to expect under the new. It matters little if a man have passed a lifetime under the old system. He must judge the new system by the experiences developed under it, and not by the old.

THE NEW SYSTEM LIKELY TO BECOME A PRECEDENT.

It is of very great importance, while these changes are passing, to call attention in time to the high duty of setting right precedents in the new system. Let the first twenty years of the new order of things, or the first ten, be managed carelessly, and the needle will be threaded wrong for fifty years, and will not be threaded wholly aright for a hundred. A responsibility of an extent and weight not easily overestimated, rests upon the manufacturing and operative classes who are now organizing a completely new factory system for the largest trade of the nation.

LYNN LEADS THE TRADE.

3. This city leads this largest trade of the United States, which is now in a condition of transition to a new system; and the fashions which shall be set here for that new system have more than a local importance.

In 1860 Philadelphia led this trade. But Lynn has now led it for ten years. The largest business of any one town, in 1860, was that of Philadelphia, in which the production amounted to $5,329,887. Next came Lynn, in which the goods were valued at $4,867,399. Then followed Haverhill, with a production of $4,130,500. Fourth, came New York, in which the products of the trade were valued at $3,869,068.[1]

But Lynn has surpassed every other city in this branch of manufactures by the rapidity with which her factories have introduced the new machinery. The patent on the McKay machine secures to the proprietors the payment of a small royalty for every pair of shoes manufactured by the aid of

[1] Eighth Census, Preliminary Report, p. 69.

this invention. It is of extreme interest to notice, by figures that are already open to the inspection of the trade, that of all the shoes sewed upon the McKay sewing machines in the whole country, in 1870, twenty-eight per cent came from this city alone. In 1870, the amount paid in royalties to the proprietors of this machine was, for the whole Union, $400,011.08. Of this sum there was paid from this city $112,924.48, or more than a quarter of the whole.

Suppose that this city led the coal trade, or the iron. You would have a responsible position. Suppose it led the woollen or the cotton. You would have a position more responsible. But your present position is more responsible than even the latter. Suppose that this city led either the coal, the iron, the woollen, or the cotton trade, and that new inventions were to revolutionize the processes of industry in that trade and that you were called to adopt a new system. Your present position is more responsible still.

DISTINCTION BETWEEN THE FLUCTUATING AND THE UNINTERRUPTED INDUSTRIES.

4. In the article produced by the shoe trade there are, and in those produced by the coal, iron, woollen, and cotton trades there are not, annual fluctuations of fashion; hence the former trade is, and the latter are not, subject to annual fluctuations of activity; and, for this and other reasons, the shoe towns and the shoe factory system are not parallels of the cotton towns and the cotton factory system, and the new system cannot meet its exigencies by copying the methods of the old.

I beg leave to make a distinction between the fluctuating and the uninterrupted industries. More than one problem is explained for the student of the high theme of the moral and industrial economy of cities, by this distinction.

Certain trades produce articles in the very nature of which there are constant and wide changes of fashion. Evidently these articles cannot be accumulated in advance, for the fashions cannot be foreseen at any great distance. A stock of outgrown fashions on the market might ruin these trades. As soon as certain annual fashions are set for the articles, these industries have a period of extraordinary activity. When the demand is supplied, a period of comparative inactivity follows until the next set of fashions is determined. If fashions fluctuate annually, these trades fluctuate annually. If fashions fluctuate twice annually, these trades fluctuate twice annually. On the other hand, it is evident that if a trade produces an article in the very nature of which there does not exist this susceptibility to a change of fashion, it may work from year's end to year's end, and accumulate, if need be, a stock of its own products.

The latter is the condition of the coal, iron, woollen, and cotton trades. The former is the condition of the shoe trade.

All trades producing articles of clothing are subject in large towns to vast annual fluctuations of activity. In Boston, for example, the length of the working season for tailors and tailoresses is estimated at ten months; for shop work, at ten; for paper collar makers, at ten; for hosiery and rubber and elastic goods, at ten; for hatters, at eight; for corset makers and hoop-skirt makers, at seven and a half; and for straw workers, at seven. It seems a mystery that so many workmen, worthy in every way, and sure to find difficulty or distress because unable to obtain occupation elsewhere, are dropped mercilessly from these employments by the thousands, at certain periods of the year. The explanation is simply that these employments produce articles subject to wide, annual, and unforeseen changes of fashion, and cannot accumulate stock in advance that is likely to be outgrown. We are often comfortably told that the wages given in such employments are of fabulous rates by the day or week. This is not often the case; but, even if it were, for how many weeks in the year does the working season hold?

There is another class of fluctuating industries in which the variations of activity arise from the changes of the seasons. Thus the length of the year is estimated for quarry workmen at ten months ; for farm laborers, at eight ; for masons, painters, and plasterers, at eight ; for brickmakers, at seven.

Now it happens that the largest trade of the United States produces an article notoriously subject, especially in the varieties requiring the most delicate and skillful work, to wide changes of fashions ; and those fashions themselves are largely affected by the changes of the seasons. Both the great causes, therefore, which produce variations in the fluctuating as distinct from the uninterrupted industries, powerfully affect the largest trade of the United States.

Lynn works but about ten months a year. Lawrence and Lowell work twelve.

This indicates the most fundamental distinction between the large cotton towns and the large shoe towns. Many give it as their judgment that the working season here does not exceed nine months.

You will mend these lulls, you say ? Hundreds of years the artisans in other fluctuating industries which I have just named have tried to mend the lulls in large towns in their trades. They have not succeeded. To do so would be to counteract a natural law. Not one only but each of the great natural causes of fluctuation so affects the shoe trade in large towns, that I see no prospect of the lulls in it being soon removed ; neither do any of the manufacturers with whom I have conversed. Rapidity of production being one of the causes of the lulls, it is found that as machinery becomes more perfect, the working season tends to become shorter. Machinery grows more perfect every day. It is introduced into large towns more promptly and abundantly than into small.

In a city establishment containing, for example, operatives enough to produce twenty sets, or twelve hundred pairs of shoes a day, the manager gives out stock enough in the morning to make only twelve or fifteen sets. As the brisk season of work arrives, stock enough is given out to make thirty or thirty-five sets a day, and more help engaged if it can be found. But, as the season of inactivity cames on, the stock is diminished again. Perhaps with a working capacity of twenty sets a day, only enough stock will be given out for twelve or ten sets. Of course, workmen drag on without half enough work for a while; and, finally, are unoccupied by the thousands.

Precisely here arise the chief industrial perils of the operative class in this branch of manufactures. Precisely here is the origin of large floating populations, with their attendant startling moral perils.

PERILS OF CONGREGATED LABOR IN LARGE TOWNS.

5. Congregated labor and a large floating population are historically known as having always heretofore given rise in large towns to grave moral and industrial perils and abuses; and the new system of the shoe trade necessitates congregated labor; and the annual fluctuations of the activity of the trade give rise in large towns to a large floating population.

ENGLISH FACTORY LEGISLATION AN ILLUSTRATION OF THESE PERILS.

Sir Robert Peel told the English Parliament in 1816 that unless the tendency of congregated labor under the factory system in large towns to give rise to perils and abuses, could be corrected by decisive legislation, the great mechanical inventions, which were the glory of the age, would be a curse rather than a blessing to the country.[1] "These were strong words from a master manufacturer," says the Duke of Argyll, writing in 1866, "but they were not more strong than true."

It is of high interest to notice that almost precisely one hundred years ago, the cotton factory system, on account of new mechanical inventions, was passing through a great transition exceedingly similar to that which the shoe factory system is now passing from the same cause. One hundred years ago this year, Sir Richard Arkwright perfected that marvelous combination of mechanical adjustments known as the

[1] Hansard's Parliamentary Debates, Vols. xxxi. and xxxiii., Sir Robert Peel's Speech on Motion for a Committee, April 3, 1816.

spinning frame. Hargreaves' great invention of the spinning jenny took place in 1765. And Crompton's celebrated combination in the mule jenny of the two preceding machines, followed in 1787. In strict analagy with what is now passing before our eyes in the history of a great sister industry, the invention of new machinery in the cotton manufacture revolutionized its processes; and the invention of one important machine necessitated the invention of others.

But the steam engine had not yet appeared. A factory system therefore sprung up in connection with vast establishments located on streams. Of necessity, the sites chosen were, in a majority of instances, at a distance from preëxisting towns and in thinly populated districts. In order to secure permanent labor, a system of apprenticeship was adopted, by which operatives were bound to work for a definite period. The consequences of congregated labor under no regulation except the unrestrained competition of manufacturers, began to appear. Hardly more frightful abuses have sprung up under the factory system in large towns than sprung up in this first factory system outside of large towns. A whole generation of boys and girls and youths and men and women of all ages, says one of the most considerate of historians, "were growing up under conditions of physical degeneracy, of mental ignorance, and of moral corruption." The very title of the bill by which Sir Robert Peel began, in 1802, the great series of the English Parliamentary Acts in promotion of factory reform, was, "For the preservation of the health and morals of apprentices and others employed in the cotton and other mills, and in cotton and other factories." The health and morals! Upon these points all the vast mass of English factory legislation turns to the present moment. It is significant to notice that when congregated labor under the factory system was tried for half a century in England at a distance from large towns, it exhibited, taken by itself and aside from any now outgrown evils of the plan of apprenticeship, a tendency to perils and abuses such as to call for the most decisive parliamentary interference.

The new star of the steam engine blazed across the mechanical sky; took a fixed place in it; and immediately there was a new grouping of constellations. The vast manufacturing establishments which existed at a distance from towns were transferred to crowded populations. Between 1802 and 1815, the factory system was transformed into its present shape. It was the birth of the inventions of Hargreaves and Arkwright and Crompton and Watt. It was a system wholly new in the world. Immediately, a tendency to perils and abuses appeared which called for vigorous parliamentary repression. English Parliaments have not been remarkable for unnecessary interference with trade, nor for sentimental legislation. The larger part of the manufacturing wealth of the kingdom was thrown into the scale against factory reform. But the cause of that reform has steadily advanced because Parliament has been forced by the terrible revelations of its own commissions of factory inquiry, again and again to interfere. The moral and industrial perils of congregated labor under the factory system in large towns! It was thought that the tendency of the factory system to these perils was corrected by the great Factory Act of 1833. Eleven years passed. The Factory Regulation Act of 1844 was found necessary. Two years ensued. Interference, always unwelcome to Parliament and always against some of the deepest traditions of English law, was found needful in spite of previous interference. In 1847 the celebrated Ten Hours Act was passed. Experience continues to teach. In 1853, the Childrens' Labor Act is found indispensable. Against every one of these great measures, the larger part of the leading manufacturers threw their heaviest influence. I recite before this assembly the list of the great Acts of factory reform wrung from Parliament, in Great Britain, to prove the inherent tendencies of congregated labor under the factory system in large towns to moral and industrial perils and abuses. A board of factory inspectors, with ample powers, sits to-day in London, with subordinate inspectors in various districts making reports to the central officers weekly. The number of prosecutions and informations instituted by the inspectors under the factory acts from 1836 to 1854, was 3696.[1]

This is the historical reputation of that system of congregated labor in large towns into which the largest trade of the United States is now inevitably passing.

LARGE FLOATING POPULATIONS OF TOWNS ENGAGED IN THE FLUCTUATING INDUSTRIES.

Ominous enough in itself, this fact is yet more ominous from the most important circumstance that this vast branch of manufactures belongs to the fluctuating rather than to the uninterrupted industries; and must, on that account, give rise in large towns to large fluctuating populations. The perils of congregated labor in large towns are large enough; but the perils of congregated labor in large towns with large floating populations have an established name that makes it impossible to speak too strongly of the worth of family life as a moral police in society.

He who comes home at night to a circle that know him well and watch his daily course, has a kind of daily appearance to make before a moral tribunal. The bliss of the home affections is a shield from vice, not only because it is bliss, but because it makes any conduct that needs concealment from the moral tribunal of the most intimate circle as painful as the bliss of ingenuousness and trust is great.

From side to side of the globe every place where a large floating population congregates is found to be a stormy moral coast. In face of universal experience I need not pause to prove the moral perils of homelessness. Those centres in New England where large floating populations gather will always be found to exhibit peculiar moral perils.

All the more to be honored and trusted for their endurance of the breakers, is that percentage of most worthy people to be found in every floating population. Not only am I aware of the existence of hundreds of excellent people in floating populations, but also of the duty of receiving these with

[1] Encyclopedia Britannica, Eighth Edition, Vol. xxi., p. 791.

especial cordiality to our hearts and homes. But in a large town, there is in a floating population not only an intermixture of the thoughtless and giddy and falling, but, further down, and most to be feared, a percentage of the thoroughly bad. Men and women who have the worst of reasons for leading a floating life need not be many in any floating population to do immense mischief. New England is not so saintly in her cities that she can afford to forget that the exigencies of trade and the wonderful growth of means of intercommunication, have brought into some of her inland large towns evils thoroughly analagous to the old and traditional evils of seaports. All kinds of people gather in a floating population. In a large city, in a floating population, it is not incautious to ask, not every tenth man, but every tenth man who pretends to a peculiar interest in your affairs, Have you ever been in jail? Every great city is a collection of camps. He who knows one stratum of the society only, does not know the city. He who knows dissipated Paris does not know Paris, but only a particular camp in Paris. So of New York and London and Berlin, and every lesser town in its proportion. The moral perils of homelessness added to the perils of this bad percentage from outside, put the solemn duty upon the resident population of these stormy moral coasts to throw the moral lighthouses of church, library, and school, but especially the lighthouses of right industrial arrangements, far out upon the edges of the reefs.

But, besides the operation of the fluctuating character of the shoe trade to produce large floating populations in large towns devoted to that trade, another cause operating to the same end exists in the great subdivision of labor which the new system of the industry has introduced. A person arrives here fresh from a New Hampshire farm and never having seen a factory. It is possible for him to be taught in a very few days to perform some one of the simpler parts of the work. I have seen in the factories machines which I think I could myself be taught in three days how to feed. The result of all this is, that in the two brisk periods of your year, among the thousands who come here, there are some hundreds who expect to learn their parts on arriving. I call attention to the fact that in the new system of the trade, there are many branches of work into and out of which a man can float, without floating in and out through the gate of an apprenticeship of seven years.

So important in this city is the distinction between the floating and the resident population, that you will allow me to use the phrase floating Lynn to characterize the former, and old Lynn to describe the latter. In the fifteen thousand now employed here in your chief industry, which is advancing so rapidly in size, there is already a population of five or seven thousand, who are here or not here, according as your business is at its points of greatest or least activity. I speak to-night with the question constantly in my mind how large that floating population is to be in ten years, in twenty, in fifty, or a hundred, in this city, in Haverhill, in New England as a whole, in New York, and in Philadelphia.

CAUTIONS IN RESPECT TO THE DISCUSSION.

The hour that is passing before this assembly is a serious one, for we are called now, in the face of these five considerations, to study our duty in respect to the special measures of prevention and reform, which are required by the solemn and multitudinous voices of local and national interests concerned in the perils which history and the most recent experience point out as undeniably involved in the very nature of the largest trade of the United States.

It is little for me to say, speaking on a theme like this, that I will not be the instrument of the capitalists. I will not be the instrument of the manufacturers. I will not be the instrument of the Crispin Lodges. I will not be the instrument of the parlors of any church. I will speak to-night in no sense as the apologist of this class or that. The theme is too serious for partisanship, or for the concealment of truth.

Let me say also that I have not suffered myself to take up

a theme so complicated and weighty without an extended and most serious attention to it, not as exhibited in books merely, but as seen in the swarming life of this city; not as seen in the opinions of this class of men or of that, but as seen by men who have the most different interests involved concerning it, and the most widely separated points of view. I have been through more than a few of your factories. I have conversed with a large number of your leading manufacturers. I have consulted carefully with many working men.

You will not allow yourselves to be confused by the criticisms I make on floating Lynn, which I have already distinguished from the resident population of the city.

The chief proposition I defend is that the working class of the manufacturing centres of New England have a right to ask of the employing class, that the moral perils of the workrooms under the factory system, shall be made, for themselves and for their children, as few and small as possible.

FOUL AND CLEAN SYSTEMS OF WORK-ROOM MANAGEMENT.

There is a foul and there is a clean system of workroom management in shoe factories. To speak at once to the point, there are workrooms in this city, in which men and women, boys and girls, gathered in large part at random out of a floating population, are sandwiched together like herrings in a box; and, uninterrupted by the noise of machinery, it is not infrequently foul talk, profanity, and tobacco smoke from morning to night! I am not speaking of cotton factories, in which the noise of machinery prevents free conversation between operatives, and in which I should not call for a separation of the sexes in the workrooms. But in shoe factories it is notoriously easy for a few foul mouths, not hard to be found in a floating population, to corrupt a whole room. This herring-box system I call a foul system. (Applause). I ask you and I beg the general public to notice that I am venturing all these assertions before a crowded audience which understands perfectly, from observation close at hand, and better than any other audience in New England, the facts of which I speak. There is not the slightest business necessity for mingling the sexes in the workrooms. But, besides seeing the sexes of about the same numbers in closely packed rooms, I have sometimes seen four or five young women crowded into the same room with twenty-five or thirty men ; or three working thus ; or two ; or one. I do not assert that a majority of mouths are foul in the factories ; but I deliberately make myself responsible for the public assertion that a father who wishes the welfare of his daughter cannot be expected to put her into factory life in a large proportion of the workrooms of this city. There is no saying more common here than that a father does not like to put his daughter or son into many of the factories. The common and permanent opinion as to what the answer would be to the question, Would you put your own daughter into workrooms managed on such a system ? is a test of the character of that system. A management in respect to which the answer to this question is notoriously and always No, I call a foul system. Perhaps I have put more than a hundred times this question, or its equivalent, and have been answered invariably in exactly these words, or their equivalent: "Before putting my daughter into workrooms managed on that system, I would see her, in some other place, work her fingers to the bone!" This is a terrible condemnation of a system wholly unnecessary in itself; affecting, here and elsewhere, a vast operative population ; and likely to affect a population larger and larger. (Applause).

On the other hand, as the example of five or six of the largest factories here abundantly proves, there is a clean system of workroom management in shoe towns. In one of the largest factories of this city I have seen the sexes in separate rooms everywhere from basement to roof. Where this arrangement is made and care is taken to appoint men of irreproachable character to oversee the workrooms of the men, and women of irreproachable character to oversee the workrooms of the women, the answer to the test question is different. I have information as to single rooms in this city in which there is every reason to believe the moral condition is good, because care has been taken as to the moral character of overseers; and, as to others, in which there is every reason to believe the moral character is bad, because there has been carelessness as to the moral character of overseers.

When the character of a floating population, the effect of the floating on the resident population, the inflamability of human nature, the immense numbers likely to be affected by the varied influences of the workroom arrangements, are kept in view, all that can be said in respect to the foul system is simply that capitalists and manufacturers ought to have sense enough not to adopt it. One hardly feels like offering arguments in the case. It is, however, as a temporary arrangement, though not as a permanent, slightly cheaper to manage on the careless system than on the careful. There is, too, now and then a man of theory, or some

"Lily-handed, snow-banded, dilettante"

critic knowing nothing of manufactures, who, overlooking the immense distinctions between the influences of the sexes on each other in the parlors of good society, or in a high school, for example, and their influences on each other in these rooms, filled from a floating population without any careful sifting of characters at the doors, judges on general principles, without having examined the case in actual life, that the mingling of the sexes in these workrooms from morning to night may be an excellent thing. And there are others, who, judging from some exceptional instance or instances, where the character of those engaged in particular rooms has been particularly good, and the overseers men of irreproachable character, and the sexes mingled to apparent advantage, think that this is the best *general* rule for the large, floating populations of the manufacturing centres of this trade, present and future, in New England and elsewhere. I fully admit the existence of such exceptions. No one is prouder than I am of the record of the Lowell factory girls in publishing the celebrated *Lowell Offering*, made up exclusively of their own contributions. There are, to my personal knowledge, workrooms in this city which may take rank in every respect with those from which this paper was issued. But it is amazing that any man should forget the immense changes in the general character of our operative populations in the last twenty-five years. Is it possible that there is a man in this audience into whose ears it never has been whispered that the operative populations in New England are not quite what they were twenty years ago? The suspension of the *Lowell Offering* in 1848 marks the beginning of a change which has not left behind it the hope that such publications will be numerous in the future. All these facts are matters of public notoriety, and not to be answered by the recitation of any score of exceptional cases.

For the sake, therefore, of any who have judged the matter at a distance, or from exceptional cases, I purpose to name a few arguments not needed by any man who will examine the theme close at hand, or by a wide induction of instances.

ARGUMENTS AGAINST THE FOUL SYSTEM.

As definitions, let me state, once for all, that by the foul system I mean the mingling of the sexes in the workrooms, and carelessness as to the moral character of overseers; and by the clean system, the separation of the sexes in the workrooms, and carefulness as to the moral character of overseers. My main assertion is that the moral perils of the latter system are less than those of the former. By workrooms, I mean the rooms which may be strictly so called. Of course I have no objection to lady book-keepers: they are usually more steady than young men. Nor have I objection to female help in the packing rooms. In these the proprietors themselves are usually present.

I shall give ten reasons for two special measures, the separation of the sexes in the workrooms and carefulness as to the moral character of overseers: ten reasons for two things; and, as I beg you to notice, for the one measure as much as for the other.

THE DOOR OF ENTRANCE TO THE WORKROOMS, NOT A SIEVE.

1. The chances in the shoe trade in any large town are extraordinarily great that bad men and bad women will occasionally be found in the workrooms; and these chances arise from the five circumstances, (1) That the door of entrance to the workrooms is not, and, on account of the number of changeable operatives, is not likely soon to be made, a moral sifting machine; (2) That the industry has each year two brisk and often painfully hurried periods, and two of comparative inactivity; (3) That the percentage of operatives changeable within the year is large on account of these fluctuations, and is estimated here to be thirty-three per cent of the whole number; (4) That, on account of the fluctuations of the industry, the floating population of shoe towns is likely to be large, and it is out of this population, itself not sifted, that operatives, in the hurried periods of the work, are taken into the workrooms through a door that is not a sieve; (5) That it is notorious that in the exigencies of the trade in the greatly hurried seasons, operatives that are expert are often retained for their expertness without any high degree of caution as to their moral character.

Such, in a large town, is the very structure of this branch of manufactures. The very nature of the industry points by all these five arguments to the necessity of a separation of the sexes in the workrooms, and carefulness as to the moral character of overseers. If there is a business necessity for packing inks and silks together, it is business sagacity to pack them as separate parcels.

All the factory streets of this city just now blossom with advertising boards for operatives. A strip of lasting board is put out at the door or window of an establishment with the words on it: "Heelers wanted"; "Stitchers wanted"; or, "Binders wanted." At two periods of the year these lasting boards are one of the most characteristic sights in your city. You know perfectly well that, when work is greatly hurried, the first workman that comes that is skillful is likely to be taken for the time. At Lawrence and Lowell operatives are engaged, in a majority of cases, by the year. They sign formal regulations on entering a factory. Neither of these customs exists here. The first argument for the separation of the sexes and the appointment of only men of irreproachable character as overseers in the workrooms is, that the lasting board is not a sieve.

OPINIONS OF RESIDENT PHYSICIANS.

2. Allow me to say that there is no class of men whose opinions I respect more highly in regard to the real inner life of cities, than I do that of such physicians as combine among their qualifications scholarship, experience, ability, and candor. It does no harm for the student of the moral condition of cities to ask questions of such men. While I continue to be such a student, I shall keep my ears open to this testimony, if my ears are not long ears.

I have put behind me all that I have now said from the beginning, in order that I might fitly, as I now may, state the decisive argument that physicians in this city, resident here for a long period, and combining the qualifications of sholarship, ability, and candor, lay at the door of the new factory system, or want of system, a startling increase of vice in the floating population. It is, of course, difficult to make estimates on such a subject. Even when they are numerous enough to nibble at any careless feet on the pavement, and to rustle now and then in the outer walls of private homes and of churches, as God grant they may never be here, there is no census taken of gutter rats. He is somewhat more than simple who looks at the police record for a complete account of all that passes in a large town. The most solemn public responsibilities make it my duty to know this city thoroughly. There is one question which tests the secret trend of under currents in a crowded population very well; and that question I have put. Of a physician having all the qualifications I have named, I asked what in his judgment had been the growth of the worst vices here in the last ten years, as tested by their physical penalties coming under medical observation. In two conversations, he told me that his judgment was, that, in the floating population of this city, the infamous diseases had increased ten per cent in ten years in proportion to the population! I was moved as if smitten by an electric bolt. He adhered to his opinion; and, after making all allowance for the war, and every other cause, laid the horrible blame upon the mingling of the sexes in the new factory system of this city. I thought the opinion worth testing by inquiry. I wished to see whether others agreed with it. You will notice that it referred to floating Lynn, and by no means to the resident population. Riding, soon after I received this opinion, with two men advanced in years, but of excellent ability and who had known the city intimately from their boyhood, I quoted the opinion and asked their judgment as to it. They both said, without hesitation, that they should not wonder if the opinion were just, such were the moral perils of a large floating population. I had occasion not long after to call at the office of a gentleman whose public position gives him an excellent knowledge of the city and especially of the wants of the working people. I mentioned the estimate to him. He said at once that it was not too high, and perhaps not high enough. A group of gentlemen who sat by seemed to give a general assent to his opinion. Startled beyond measure, I selected a second physician whom I considered the equal of any other in the city in the four qualifications I have named, a man whose balance of mind you all honor, but whom I must not more particularly describe; and, calling on him one evening, was most cordially received. I simply said that I was investigating the moral condition of the city; that I valued the aid of a physician's judgment; and that such and such an opinion had been given to me, mentioning both the estimate I have already quoted, and the name of its author, though saying nothing whatever of the factories. "The estimate is not high enough," was the reply. "Since the sexes have been mingled in the new system of the factories, the floating population has shown the effect. And what less could be expected ? Take a room twenty-five or thirty by fifteen or twenty feet in size ; put a dozen men and a dozen women, and boys and girls into it; and let tobacco smoke, and profanity, and foul talk take their course. The estimate is not extravagant, if it is high enough, in respect to the effect on the floating population of the mingling of the sexes in the factories." Before this opinion was given, not only had I myself not said a word of the new system ; but, so far as I am aware, the physician was not informed from any other source that I was at all interested in the topic. The reference to the factories on the part of this able physician was wholly gratuitous and unprompted. " But," said I, " what is the effect on the resident population of these perils in the floating ? It is hard to make an estimate ; but, at Nahant, the people tell me that, after a storm, it is necessary to wash slight salt stains from their windows, so much of the spray is blown across the peninsula. Does this floating Lynn, foaming against the resident population, dash any spray over it ? " " The figure is a very good one," said the physician : " two per cent of the resident population may be injured." I afterwards consulted carefully with a third medical authority, having all the qualifications ; and he endorsed fully, and with minute details of facts, the opinions of the other two, both as to the extent of the evil and as to its cause !

3. With the eyes with which you are looking into my face at this moment, I need only to name, as a third argument, the possible future effect of the floating population on the resident, when the former is larger. When the operative's fever broke out into the middle classes in Old England under the old factory system, Robert Peel led a reform ; and the abuses in the mills that had been overlooked while only the operatives suffered, were remedied almost exclusively from fear of contagion in other classes.

4. Nor do I need do more than name, after this apalling testimony, the topic of foul mouths in factories.

5. It is plainly impossible to make a proper provision against

these evils by either of the two measures taken without the other. Some two or three months since, I visited a factory room on especial invitation. The person who invited me and who showed me the room, had spoken of one young woman in it as an exception to the rest, and as resisting every unhealthful moral influence. But he told me that at a previous time the room had been so remarkable for foul talk that no young man could work there any long period and retain his virtue, the girls were so bad. I saw a room some seventy by twenty feet in size; and several young men in one end and young women in the other; cutters and stitchers; no connection of their work requiring that it should be in the same place. When in the room I asked my informant if a particular one of the young women was not the exception of which he had spoken. My informant said she was. The others were coarse looking. I instance this case in order to add that in any room arranged as that was, even if Washington or Lincoln were overseer, he could not prevent the moral condition from being unhealthful if the character of those in the room should happen to be bad; nor, on the latter supposition, would separate rooms without good overseers be enough. Neither of the two measures will be found sufficient alone to meet the exigencies of manufacturing centres.

6. It is found by experience that it is in the workrooms that a young woman coming here and not resisting, as, thank God, hundreds and thousands do resist, the morally unhealthful influences, looses that natural shyness and modesty which are her charm, and gradually acquires a repulsive boldness. There are spiritual and physical signs for every vice. The loss of spiritual shyness and nobleness can be seen, as well as the loss of natural freshness of complexion and of a lustreful flash of the eye. Suppose that a young woman coming here for the first time falls into both an illy regulated boarding-house and a room of unhealthful moral conditions in a factory. Which will do the more harm? Which will begin the harm? Where will the first indentation of ill occur? Evidently she can choose her companions to a great extent in the boarding-house;

and, if she is of high principle, will choose the best she can. But she cannot choose her company in the workroom. She must breathe the atmosphere of the company in the latter eight or ten hours a day. She may, in a large measure, choose her own company in the former, except for perhaps an hour a day. Further on in the history of deterioration, the illy regulated boarding-house and the street school may strip the flesh from the peach, but the down of the peach was brushed away in the workrooms. This is found to be the history of the case in tracing almost any individual example of deterioration.

7. It is plain that neither boarding-houses nor churches can do as much for a floating as for a resident population. If the floating population of this city had homes here, I have no doubt it would be as free from moral perils as any population of its size. Not only have they no homes, but they often have difficulty in finding boarding-houses under the best management. No system of corporation boarding-houses, such as exists in the cotton towns, exists here, chiefly because the occupants would most of them be out of the city three months in every twelve. In a large town devoted to your branch of industry, many things cannot be done by boarding-houses and churches which can be done in a cotton town; and, on this account, what can be done by right industrial arrangements is of all the more urgent importance.

8. I feel it a descent to mention any financial argument; but, if any business man doubts that whatever undermines good order in the working class, also undermines regularity of industry and diminishes the worth of the hours of labor, he need not look far into the history of the factory system to find an opinion not weakened by a doubt.

9. The sentiment of working men and women of the best class is eager and emphatic in favor of the separation of the sexes in shoe factories. I have found this true on all hands in my conversations. Obviously the working people are, in this matter, not only the chief sufferers. They are also the best informed. The employers of the present day, though

they may have had personal experience as workmen under the old system, have usually, if they are past middle life, had none as operatives under the new. It is worth little for an employer to say that he has had thirty or thirty-five years experience in the shoe business. There have been hardly more than ten years experience in the shoe business in its present form. An employer without experience as an operative, and who judges of what happens in his establishment by what occurs in his presence, is not as wise an adviser in this case as the operative. The sentiment of the working men and working women of the best class is the decisive consideration.

10. The sentiment and the practice of capitalists and manufacturers of the best class are to take pains to appoint men of irreproachable character as overseers, and to separate the sexes in the workrooms. On several occasions, after bringing this topic privately to the notice of manufacturers, as I have repeatedly done, I have found them condemning the foul system, although, on account of some temporary inconvenience in the structure of their factories, their own establishments were not yet wholly free from it. In spite of the practice of the majority of the factories of this city, all of perhaps a score of manufacturers to whom I have presented the subject privately, have, except one, unequivocally favored the system of the separation of the sexes. I must not leave the impression that I have addressed you to-night by the prompting of any one, for I have spoken wholly as an independent student and critic of public affairs. Let me state, however, that I shall not soon forget the emphasis with which, at the head-quarters of this industry in Pearl Street, Boston, one of the ablest merchants there, himself owner of large factories and member of a factory firm known and honored throughout Eastern Massachusetts, said to me, on my bringing the topic of to-night to his attention, his own practice corresponding with his advice: "Do you adhere to the ground of the separation of the sexes. It is of high public importance. There is not the slightest business reason for mingling the sexes in the new system, and any number of moral and industrial reasons against it."

Why discuss this subject publicly? Because only a powerful public sentiment will correct the evil. In what method will public sentiment aid? It is not difficult to point out the steps. Let it be made socially as unpopular for a man to manage a factory on a careless system and mutilate souls as to manage a railway on a careless system and mutilate bodies. Then the better class of men will be influenced. Let a majority, thus gradually won, set right fashions, and even the moneygripes, and men lower down, will be reached. Business is a regiment. For industrial reasons men must keep step with each other in it. Let a majority of the Board of Trade of any city set right business fashions, and the inferior men who care only for money are usually brought sooner or later to respect the step of the regiment.

CONCLUSION.

I need only to invoke the visible presence before this assembly of the lofty spirits of Sir Robert Peel and Lord Shaftesbury, to suggest sufficiently the historic perils of congregated labor under the factory system in large towns. Would that in the air above every manufacturing centre of New England, Robert Peel and Lord Shaftesbury, colossal and admonitory in archangelic stature, might each stand to teach, with one hand pointing toward Old England and the other stretched as a shield over New England, the methods of avoiding here the perils which have arisen there.

God grant that the day may never come when American society shall be divided into two classes, the unemployed rich and the unemployed poor, the former a handful and the latter a host! To that we shall come, as so many parts of Europe have come already, when our population is as thick as theirs, unless all good men and true unite to keep certain industrial ghosts that now trouble Europe from crossing the Atlantic.

Two great axioms rule modern manufactures. They are, that subdivision of labor increases the skill of the workman; and that, other things being equal, the larger a manufacturing establisment, the greater the profits. These are the organizing laws which explain most of the phenomena of manufacturing populations; and will continue to explain them for ages to come, although it is only in the last age that the laws can be said to have been discovered.

It is the principle of subdivision of labor which confines the modern operative more and more to some single detail, the work upon which, after it becomes a habit, calls into activity only a few of the mental powers; has in it no variety, and so does not develop the mind by tasking it at different points; is in itself of only petty importance, and so excites little enthusiasm in labor and even little pride of skill. De Tocqueville, in a celebrated passage, discussing the modern science of manufactures, asks what can be expected of the human intelligence, when, year after year, for twelve or ten hours a day, it is occupied in the single detail of making heads for pins.[1] The principle of subdivision of labor has an inherent tendency to dwarf the operative mind, unless the most powerful stimulants are applied outside of factory hours to develop the faculties which the manufacturing work never calls into activity. Outside of factory hours! Those words are lightly uttered only by the inexperienced in operative life. Outside of factory hours there are, properly speaking, for operative populations tasked ten or twelve hours a day in close apartments, no hours at all. The labor of the mill or of the mine, which goes on in all weathers with the invariability of the sun in its courses, is not to be compared with agricultural labor, interrupted by the changes of the seasons and even of the daily sky. Twelve hours or ten in a factory, and then three hours or two enthusiastic pursuit of mental culture! No eyes yet born are destined to see that wonder grow common. There are a few mental and physical constitutions vigorous enough to combine these two sets of hours, and so counteract the narrowing mental effect of labor for years at one unvaried mechanical detail. But the mass of operative populations can be expected to exhibit no such physical, to say nothing of such mental and moral, vigor. They are swept remorselessly

[1] De Tocqueville, Alexis, Democracy in America, Vol ii., Book ii., chap. xx.

under the wheels of subdivision of labor and long hours. In women and children, who constitute nearly half of operative populations, how much life is left for mental culture after twelve hours severe labor in a mill? But subdivision of labor increases skill; increase of skill increases productiveness; increase of productiveness increases profits; and long hours are the scythes that reap the gain. This is the law of manufactures; and, it is only saying what is evident in the nature of things, and no less evident in the condition of all manufacturing populations where factory occupation has been hereditary for three or four generations, that the tendency of the system is to make the operative class inferior; and the inferior yet more inferior. Emerson stood at the door of the factories of Great Britain and wrote that society is to be admonished of the mischief of the division of labor by the fact that, in three generations, the robust, rural Saxon had degenerated in the mills to the Leicester Stockinger and to the imbecile Manchester Spinner, far on the way to be spiders and needles.[1]

On the other hand, the operation of the principle that, other things being equal, the larger a manufacturing establishment the greater the profits, tends to call out all the capabilities of the minds that lead and organize in manufactures. It requires capacity in that class, attracts capacity, and tasks capacity. Men of education are often drawn into manufactures by the allurement of the size of the enterprizes involved. The tension of mind, and the variety of its applications in the conductor of a large establishment are at all points a contrast with the condition of the mind of the operative.

Inevitably, therefore, as the effect must follow the cause, the system of modern manufactures in large populations tends to produce a superior class and an inferior; and, as years go on, and the first effects themselves become causes, to make the superior yet more superior and the inferior more inferior. I am not denying the advantages of manufacturing eminence; but stating, as a motive for public caution, what political economists have long acknowledged as the disadvantages of such

[1] Emerson, R.W., English Traits, chap. x.

eminence. Even John Stuart Mill, using England as a lens and putting behind that telescope the best eyes of political economy, writes a deliberate chapter[1] on the Probable Future of the Laboring Classes, and goes so far as to say that he finds the prospect hopeful only because he expects the whole system of wages to be superseded by that of coöperation. But the system of wages is interwoven with the whole structure of modern life, and does not show a tendency to vanish out of history like a morning cloud. The accumulations of wealth fall chiefly to employers and not to operatives. The distance between the two classes is a result of deep causes arising from the two great laws of the manufacturing system. It is out of these laws that there inevitably originates what has been called, in modern times, a manufacturing aristocracy. De Tocqueville, using this phrase, compares the territorial aristocracy of former ages with the manufacturing aristocracy of to-day; and finds the former superior to the latter because it was bound by law, or thought itself bound by usage, as the latter is not, to come to the relief of its serving men and to succor them in their distresses.[2] I see no charm in democracy that can alter the nature of things. The subtle laws of subdivision of labor and of size of establishment apply to manufactures in New England as well as in Old England. Under some restraints from the nature of our institutions, they will, notwithstanding, produce here as there an employing class and an operative class; and perpetually tend to make the distance between rich and poor in manufacturing populations wider and wider. De Tocqueville thought that the friends of democracy should keep their eyes anxiously fixed upon the operation of these two laws; and that, if ever a permanent inequality of conditions again penetrated into the world, it might be predicted that this is the gate by which it will enter.

I find a right to speak very plain words in New England, the whole Atlantic slope of which is a factory, by lifting up

[1] Mill, John Stuart, Political Economy, Book iv., chap. vii.

[2] Democracy in America, Vol. ii., Book ii., chap. xx. Also Vol. ii., Book iv., chap. v.

my hand and pointing to Manchester and Liverpool, to Birmingham and Leeds, and London and Rouen. John Ruskin says that the unemployed poor of Great Britain are daily becoming more violently criminal. A searching distress, he told the University of Oxford last winter, invades the middle classes, arising partly from their vanity in living always up to their incomes, and partly from their folly in imagining that they can subsist in idleness upon usury.[1] Factory reform and factory legislation are old and great and grave themes in English literature. They are old and great and grave themes in Parliament. The best mark of himself Thomas Carlyle has made on the face of this planet, and a mark perhaps not soon to be erased, is his discussions of the condition of the working classes as illustrated in the problems relating to the English poor and the questions between capital and labor. Dickens has written more to illustrate this field than any other. Victor Hugo's best words are upon these themes. Romney and Aurora Leigh, in Mrs. Browning's greatest work, are typical characters for the best thought of Europe. Their chief office in the hands of the poetess is to illustrate rival methods in philanthropy toward the working classes. John Bright leads middle England because he teaches the transcendently important truth that, wherever the few live, the nation lives in the cottage. George Peabody endows colleges with his left hand, but with his right builds houses for the London poor, and schoolhouses for the poor of our Southern states. Prince Albert occupied his mature life chiefly by studying the condition of the working classes of London, and particularly the best arrangement for tenement houses. We remember what Alfred Tennyson says of this noblest of the recent English princes:

> "Laborious for her people and her poor,
> Voice in the rich dawn of an ampler day,
> Far-sighted summoner of war and waste
> To fruitful strife and rivalries of peace."

Let no one be ashamed to discuss the condition of the un-

[1] Ruskin, John, Lectures on Art before the University of Oxford, 1870, p. 27.

cleanest poor. God is not ashamed of this. At bottom this is what he discussed at Gettysburg and Richmond, at the liberation of the Russian serfs, at the granting of the Magna Charta, at the Restoration of Letters, at the Fall of the Roman empire, at Calvary itself. For two hundred years the most vital causes in politics have turned upon the condition of the many as opposed to the few. It is not likely that it will be different with history for two hundred years to come.

It is most necessary to notice that, on all these themes, the reformers and deformers are yet strangely mingled. The truth that labor reform, both here and in Europe, contains, seems to me, I must confess, a jewel in the coarsest incrustations; and yet no less a jewel for its strange setting. I suppose that the deepest question yet remaining unsolved in this nation is that concerning the relations between capital and labor. Slavery itself was but one form of that question. We hear the retreating footsteps of the discussions concerning slavery, and at the same time the advancing footsteps, under which Europe already shakes, of discussions concerning the working classes in crowded populations. The discontent of the working classes! The discontent of the working classes! This is a sound ominous of much, both good and bad; but it is not the less ominous of the one or of the other for being borne to us on all the four winds of heaven.

In one field of that wide discontent, I have spoken to-night for the largest trade of the United States. I have spoken for Lynn, for Haverhill, for all New England. I have spoken for New York and Philadelphia. Outside of the collieries of Pennsylvania, there is perhaps not a spot on the continent where I could have delivered to-night a speech more needing to be made. Go to the grave of George Peabody, almost in sight of which I speak, and learn your duties to the poor, and toward the present and the future of the vast trade which this city leads. (Applause).

LECTURE III.

SUNDAY EVENING AT MUSIC HALL.

Music Hall was packed to overflowing on Sunday evening last, it having been announced that the third of the series of meetings now being held there on Sabbath evenings, under the auspices of the First Congregational Society, would take place. The floor and galleries of the hall were completely filled at the earliest possible moment, and over one thousand persons are reported as having left, being unable to gain entrance. The excitement growing out of the turn which affairs have taken in regard to certain statements made at the previous meeting by Mr. Cook, having culminated in the appearance of the cards published in the *Reporter* of last Saturday (and which we republish by request to-day), the supposition was that Mr. Cook would allude to the subject again on Sunday night: hence the great rush to the hall. Upon making his appearance on the platform, Mr. Cook was greeted with loud demonstrations of applause. After the usual preliminary exercises, Mr. Cook proceeded to read as follows:

There is one sweet and holy game which will not be played here to-night. It is that of absurd misrepresentation of what I say. I dislike to speak from manuscript, but you will bear with me for a few moments while I read what I do not care to have turned so far end for end that white shall appear black. There is another reason why I prefer to speak from manuscript. My mood to-night is that of morning on the mountains. I feel like Bismark before Paris. I might say something incautious if I were to speak without notes here and now.

It was once my fortune, or misfortune — I hardly know which — to call on President Buchanan, at the White House, just after news of the evacuation of Fort Moultrie reached Washington. When I referred to the news, he said: "It will not do to believe all you hear. Sir, the telegraph is a wonderful invention. It bears messages that are short and weighty and rapid and pithy and false. It will not do, sir, to believe all you hear." Certain gentlemen who have not been asleep in the daytime in this city for the last six days, assure me that there passeth through the city just now a hurricane; though whither it goeth they know not; and ninety-nine per cent of those who were here last Sabbath night know not whence it cometh. Now in all such hurricanes it is endlessly important to remember President Buchanan's advice. It will not do to believe all you hear. The famous story of the tradesman in the Strand, who was told that a neighbor of his, in his illness, had thrown up three black crows, is, as I judge, rather more than outdone in our streets. This tradesman, you will remember, investigated rumor. Going back on the line of authorities for the story, he found, first, that it was only two black crows; then that it was but one black crow; and, of the only competent authority, he learned that it was not a crow at all he had spoken of, but only something as black as a crow. I advise all who are glad or sad at the flight of crows through our air in twos, threes, fours, or in flocks, to imitate the example of this famous tradesman in the Strand.

I had anticipated seven styles of sentiment as to my address of last Sabbath evening. There are three chief parties interested in the discussion — the employed, the employers, and the churches. Each of these is to be divided between the offended and the unoffended. There are six parties. But the seventh is the party of hearsay. It is this last that I anticipated would have the most wildly inaccurate impressions. It is one thing to hear a speech, and another to hear of it. It is one thing to hear with the ears, and another to hear with the elbows. In spite, however, of all disadvantages, the working men, I am assured on all sides, are with me, almost to a man. I have to thank nearly every paper of the city for a more or less favorable notice of my remarks as a whole. One of them fully agrees with me as to the moral dangers of the system, or want of system, which I discussed. I find powerful support in the churches. In estimating public sentiment, the difficulty is, that men in a particular eddy imagine the whole ocean to be moved as their eddy is.

It is my duty, as a student of public sentiment, to study all the eddies. I am assured (though I do not lean on the assurance), and assured from many very differently prejudiced authorities, that the capitalists and employers are with me, except five or six, who are very angry. I know, beyond all question, by their personal testimony to me, that every manufacturer of the many with whom I have conversed, with the exception of one, is in favor of the two measures I recommended — *the separation of the sexes in the workrooms, and the appointment of good moral men as overseers.* In view of the wildness of the statements on which the party of hearsay have formed their opinions, I think it not useless to remark that what the audience said here last Sabbath evening is, as an indication of public sentiment, not wholly unimportant. Three times during the address, and again at its close, the audience, in spite of its being Sabbath, and the place one used for the time as a church, gave open applause in a marked way. I took no pains to call out these expressions. I tried on one occasion to repress them. I have not trusted my own impressions; but have asked several gentlemen, and among them one of the best speakers of the city, what the audience meant by its expressions. I have asked if the majority of the audience meant to approve the general drift of the address. And the answer in every case has been, that the audience — which represented excellently the whole city — meant to approve the general drift.

Certain statements which I made privately in the confidence of my own parlor, have been thrust before the public in a childish card. If any one thinks that this is large and dignified business, the opinion must stand on its own merits. *I call public attention to the fact, that it was not by means of anything I said publicly that any allusion of mine in a public way was brought down to the low arena of personality.* I referred to a room about seventy feet long and twenty wide. I suppose there may be twenty or fifty such in Lynn. I have myself seen several; and it is my conviction that in some of these the moral conditions were bad, and that the moral dangers of the arrangements unnecessarily adopted by the employers were great. *I could easily have chosen illustrations elsewhere, under evidence of a different character, and fit to bear the severest tests.* I suppose that all in Lynn who have given any thorough attention to the subject know this; and the working men best of all. Of course, there are most excellent factory rooms in this city, and most excellent people in them. I could point out other rooms in Lynn where these moral dangers do not exist, or exist to a vastly less extent, and this on account of other arrangements adopted by the employers. There is a foul and there is a clean system in this largest trade of the United States; and I stand for the latter, and have defined what I mean by the former and the latter.

I do not know that any one has assailed, or attempted to assail, my chief assertion, *that the moral dangers of the latter system are less than those of the former.* The working men and working women cry out for the clean system. God hear their cry! I am their friend, and the friend of this city, and the friend of other cities having similar interests, by giving this cry utterance; and in doing so at a particular juncture, when the whole vast trade throughout the United States is, on account of the invention of new machinery, in a state of transition, and is adopting a system likely to be a precedent for five hundred years.

Old Lynn I have always distinguished from floating Lynn. I have repeatedly said that I am proud of old Lynn, although I have pointed out its dangers from floating Lynn, on account of this transition state of the great branch of industry which employs here, directly or indirectly, ten or fifteen thousand people, and for the last ten years has put this city at the head of a trade larger than the coal, the iron, or the cotton, and made the fashions it shall set for that great public interest of altogether more than local importance. Instead of the few times I have publicly touched this theme, I should have been justified, by its local, as well as by its general, importance, in touching it oftener; but, in the few times I have touched it,

I have most carefully avoided every statement that could endanger the effect of calm, guarded, and dispassionate public discussion by a pitiful descent to personalities about names and places. I made no allusion to names (as this card does, in quoting private conversation); and my whole reference was studiously intended to be so indefinite that no identification would have followed as to names and places from my language reported literally. I quoted testimony of physicians; but of this I asked in their parlors, and received from them, permission to make use. For what I have said privately, I am willing to give any man a private answer. And I shall take this occasion to say that to any generous and candid man it will be perfectly satisfactory. For what I have said publicly, I have nothing to say, except that *I* have nothing to change. The three strongest words which the card accuses me of using publicly, I never used. I never said that these persons *deserved* what was said to me of them. It is one thing to make an assertion of your own, and another to quote the assertions of another. Certain assertions were made to me about what *was* some ten months or so since, not about what *is* — the only point this evasive card covers. If what was said to me — I must say it was said — is untrue, of course I rejoice as much as any one. If, by any misunderstanding, any injustice has been done anywhere, of course I shall be delighted to see it rectified. I repeat that I could have chosen illustrations elsewhere, under different and severe evidence. But this changes no assertion of mine, and touches no important point of my argument.

The bondage of the pulpit is a theme in the thoughts of many, and on the lips of but few. I do not consider myself at liberty to allow this occasion to pass without inculcating, for the honor of the church which I represent (and it is highly to its honor), for the honor of this city, and for the public good, the lesson that Wendell Phillips is inaccurate when he sneers at the pulpits for being turned by the great wheel of the factories. This is an important public lesson; and I have friends watching me from near and from far who would never forgive me if I failed to use this occasion to utter this grave and perfectly frank word on a theme of such grave public significance. *It is really of public importance that it should be taught in some manufacturing centre of New England, and here and now as well as anywhere, that, while the envy of the rich by the poor is to be abhorred and the benefactions of the rich to the church are to be eulogized, no gold-dust is to be allowed to choke, or to bring hesitation or quaver, in the pulpit, to the clear bugle-note of truth.* The clear bugle-note of truth, I say. I mean calm, clear, dispassionate, unselfish, guarded discussion, not of what is most popular, but of what most needs to be said, even if it be a chief sin of poor against rich, or of rich against poor. The hopes of rich and poor train behind that bugle-note. I had rather my lips should be closed forever than that, on the Lord's day, in God's house, any hesitation or quaver should come into that note, with the bugle at my lips, toss into it gold-dust what hand or hands will or can. If I had any other spirit, I should be obliged to read in a sense of the profoundest irony, not intended by Mrs. Browning, her words:

> "Now press the clarion on thy *woman's* lips,
> (Love's holy kiss shall still keep consecrate,)
> And breathe the fine, keen breath along the brass,
> And blow all class walls level as Jericho's
> Past Jordan; crying from the top of souls
> To souls that here assembled on earth's flats,
> To get them to some purer eminence
> Than any hitherto beheld for clouds —
> What height we know not, but the way we know."

I had rather be any other kind of spaniel than a pulpit spaniel. If any one has mistaken me, while yet a young man, for a member of that rare species, I must remark that I hope the race is becoming extinct in New England, even in the manufacturing centres, where the distinctions between rich and poor are perhaps the widest. I may not know myself; but I do not think I am a being capable of being wrapped up in the soft, sweet-scented paper of social obligations, and put into a rich man's waistcoat pocket (or a poor man's either, for that matter, though I had much rather be in the latter than in the former). Much less am I capable of being put into the coarse paper of dictatorial and excited language, with its rough shards of inaccurate statement, inuendo, evasion, personality, and browbeating, and thrust bodily into such pocket; or pinned in, with any bar of gold, however large. I make no charges against any man. I do not refer at all to the past. I refer only to the future. But I say that if (I beg you to notice that I say *if*) any two, or three, or ten, or any other number of rich men think they own the pulpits of this manufacturing centre, I must say that for this property the great seal of the Lord the King is not on their title-deeds.

Every man to me is just as large as his capacity, his training, his conscientiousness, and his attention to the subject in hand. Take out either of these four qualifications, and the rest, however large, no matter what wealth or office adds, are not large enough to make a man large. Every man to me is just as large as his reasons. I purpose to continue perfectly open, calm, independent, guarded, public speech, with no other rules to guide me in doing so in making social distinctions, besides these.

Mr. Cook then proceeded to deliver a fine discourse upon "False and Ture Faith and Repentance," founding his remarks upon the words of the jail-keeper: "What shall I do to be saved?" He was listened to with attention throughout.

When Mr. Cook had finished his discourse, and as he was about to give out a hymn, Mr. S. M. Bubier came forward upon the stage, and asked permission to be heard in refutation of the statement made in the reverend gentleman's previous discourse. A movement being made on the part of some to leave the hall, Mr. Cook earnestly requested all to remain and hear what the gentleman had to say, and that quiet should be preserved. Mr. Bubier said he was not aware of the existence of any such place as that which had been alluded to, and that, having resided here for some fifty-five years, and done business in the city a good portion of that period, he was probably better acquainted with the system than Mr. Cook, who had only resided here seven months. Mr. Bubier here made allusion to the previous connection of Mr. Cook with an insane asylum, at which point he was interrupted by hisses and other marked tokens of disapproval. After this demonstration had ceased, he attempted to proceed, but the hissing was renewed, and he was prevailed upon to leave the stage.

Mr. Cook made no rejoinder to

the remarks of Mr. Bubier, but proceeded to read the hymn commenceing, "Majestic sweetness sits enthroned," in which the congregation were invited to join, and the audience was then dismissed, with the hope from the speaker that all would bear away with them the spirit of the hymn just sung.

We would take occasion here to enter a protest against making any demonstrations of applause or disapproval, such as were exhibited at the above meeting, upon Sunday evenings, especially when the services are of a religious character. No matter what a speaker may say, the feelings of the audience should be repressed, in respect for the sanctity of the day. It is the earnest desire of Mr. Cook that no such demonstrations may be repeated at any future service which may be held in Music Hall under his direction, as it is the design that such services shall be of a strictly religious nature. It is to be hoped for the honor of Lynn, and the respect due the holy Sabbath, that we shall not be obliged any more to record applause or hissing upon any future occasion of the kind. It is entirely out of character, and much to be condemned. — *Semi-Weekly Reporter*, Feb. 1.

A CARD. — At an informal meeting of the First Congregational Church and Society, convened at the close of divine service, Jan. 29th, the following resolution was unanimously adopted, and ordered to be printed in the several Lynn papers:

Whereas certain statements have been publicly made, in which the veracity of our acting pastor, Rev. Joseph Cook, has been called in question, and his standing with this Society as a Christian minister attacked; therefore,

Resolved, That this Society have the most entire confidence in the integrity, Christian character, and standing of our said pastor; and that we, as a Society, most heartily and cordially approve of his boldness, ability, and fidelity in presenting, both publicly and privately, those truths which, though unwelcome to some in this community, need to be plainly spoken in this day of increasing immorality and vice; and that our confidence in him as a Christian minister is unabated by the fearless manner with which he has withstood all attempts to intimidate and thwart him in his attempts thus publicly to perform an unpleasant but important Christian duty. Per order,

FRANK P. BREED,
Parish Clerk,

Lynn, Jan. 30, 1871.

A CARD. — I would hereby tender my sincere thanks to the Rev. Mr. Cook for the truly Christian stand he has taken as watchman on the walls of Zion to cry aloud and spare not those crimes in community least rebuked, and therefore the more fearful and devastating. Purity is the baptism of scientific Christianity; it should be insisted upon; and may his pulpit eloquence shatter the hoary fabric of ancient Sodom, and place mankind upon an eminence where the beams from the sun of truth first strike.

MRS. M. M. GLOVER.

TO THE PUBLIC. — Rev. Joseph Cook of Lynn, made a statement in Music Hall, during his last lecture, to this effect: That he visited a room in Lynn, from sixty to seventy feet in length, and some twenty feet in width, in which, at one end were six or eight girls employed as stitchers — at the other end were as many men. They were coarse, low, vulgar, bad-featured girls. A man who showed him the room informed him that no young man could come to that room a virtuous man and remain so any length of time, because the girls were so bad.

We, the undersigned, believing we are the persons alluded to, feel that his charges are very unjust and unchristian, and we feel justified in appealing to an honest public to investigate our characters, and see whether the investigation will warrant any person in making such charges as these against those whom we believe have nothing to condemn them but that they are compelled to labor for their own maintenance.

[*Names of eight female and of six male employees, omitted.*]

To THE PUBLIC. — I understand that the Rev. Joseph Cook claims that he got the foregoing statements from me. In the first place I never made them. They are untrue if made by any one. It *is not* the character of this room. Nor do I know of any room of which the Rev. gentleman would be justified in making any such statement. HENRY DOWNING.

The Rev. gentleman, in a conversation with us, acknowledged that the room alluded to in his lecture was one of which Messrs. Berry and Beede were the proprietors, and in that case must be the room where the signers of the above are employed.

L. B. FRAZIER,
S. M. BUBIER.

Semi-Weekly Reporter, Feb. 1.

We wish to state that the publication of Mr. Cook's vindication of his statement was not solicited by him, but requested by the press as a matter of local interest. — *Semi-Weekly Reporter*, Feb. 1.

THE MUSIC HALL MEETING.

Considerable excitement was manifested throughout the city, last week, in regard to certain statements alleged to have been made by Rev. Joseph Cook, of the First Congregational Church, in his discourse of Sunday evening, Jan. 22d, at Music Hall, upon " The Moral Perils of the Present Factory System of Lynn." On Saturday, the following card, which explains itself, appeared in the columns of the *Reporter*.

. . . . . .

The announcement by the Society under whose auspices these meetings are being held, that Mr. Cook would speak again at the same place on Sunday evening last, called out an immense crowd of people, filling every seat and standing-place in the hall, while hundreds were unable to gain admission ; it being very naturally supposed that Mr. Cook would allude to the subject that was agitating the public mind.

After making an impressive prayer, and reading a hymn, which was sung by the choir, Mr. Cook read from manuscript, as follows :

. . . . . .

At several points during the reading the audience gave evidence of their approval by suppressed applause, and at its close Mr. Cook proceeded to deliver a powerful extemporaneous discourse, illustrating faith and belief by the "parable of the iron-clad oath." The large assemblage paid strict attention throughout.

At the conclusion of Mr. Cook's discourse, and as he was about to give out the closing hymn, Mr. S. M. Bubier stepped suddenly forward from the rear of the stage, as if to address the audience. Perceiving this movement, Mr. Cook requested the audience to listen to what Mr. Bubier had to say. Waving his hand towards Mr. C., Mr. Bubier said:

"This man has made a public attack upon the virtue of Lynn. I have lived here fifty-five years; he has been here seven months, and I know not how many months outside of an insane asylum."

At this point Mr. B. was compelled to retire before a perfect storm of hisses; the audience, whatever might have been their private opinions upon the subject discussed, refusing thus emphatically to allow him to proceed further.

Mr. Cook made no response whatever; but after Mr. B. retired proceeded to read the closing hymn, which was sung; after which the audience was dismissed. — *Transcript*, Feb. 4.

The Absorbing Topic.

As we write, the all-absorbing topic in this city is the *denouement* at Music Hall last Sunday evening. The merits and demerits of the principals in that unhappy affair are freely discussed in parlors and workshops and on the streets. The unhealthy excitement caused by this affair had almost decided us to refrain from referring to the subject at all; and it is with feelings of great reluctance that we allude to it, and we only do so to satisfy the demands of an excited and expectant public. The unchristian feelings generated in society by this affair are greatly to be deplored; and however pure the motives of the principals in it may have been, the effect is unpleasant to contemplate. Although comparatively a stranger among us, the Rev. Mr. Cook has attracted much attention by his bold and fearless manner in presenting his sermons to the public; and even before we became acquainted with him we commended him for his frank, outspoken style, and for the interest which he appeared to take in the labor question and the general welfare of the masses. We were present at Music Hall, a week ago last Sunday night, and heard him make bold, astounding, and yet guarded, statements relative to the morals of those employed in the workshops of this city, in which remarks he cited the sources from which he had obtained his information, without, however, mentioning the names of persons. As has been alleged, he did then and there make certain broad and sweeping statements relative to the immoral condition of a workshop, which he described by an apparent guess at the length and breadth, and by an approximate estimate of the number of males and females employed therein, at the same time stating that personal observation coupled with the broad statements previously referred to, and which he claimed to be a quotation from the remarks of a male employee who conducted him through the establishment. Believing him to be honest in his motives, convictions, and statements, and knowing — as every observing citizen knows — that society, especially in large towns and cities, has been sadly demoralized since the breaking out of the rebellion, we were prepared to hear that our own city was not an exception to the general rule; but we frankly admit that, had not the statement about which there has been so much excitement been given as a quotation, we could not, in the absence of proofs, have done less than to attribute it to a misconception of facts or an overwrought imagination. In regard to the correctness of the other leading statements made by the speaker on the evening mentioned, we refer the public to the authorities which he cited. We left the hall that night with mingled feelings of sorrow for the sad condition of society, as portrayed by the speaker, and admiration for the man who, zealous for the cause which he has espoused, and apparently honest in his convictions and intentions, beards the lion in his den, fearless of the consequences. "To err is human; to forgive, divine." Mr. Cook, in his zeal, and by his bold, sharp, self-reliant, persevering, and determined attacks upon sin wherever found — making no distinctions between that clothed in broadcloth and fine linen and that habited in rags — has, as such men always have, aroused an opposition. There evidently has been a latent opposition to Mr. Cook ever since his advent among us.

There were, and still are, many here, as elsewhere, to whom the unveiled picture of sin in its naked hideousness strikes terror and dismay, and to them its exhibitor is an unwelcome visitor. On the Monday following the delivery of the sermon which has created such a sensation, the opposition to Mr. Cook began to assume outward form, increasing each day in certain quarters, and in the course of three days his remarks had been so distorted and enlarged that the originals were unrecognizable. We are told that on Thursday he was interviewed, in his own parlor, by two prominent shoe manufacturers, one or both of whom had been led to believe that he (Mr. Cook) had referred to their shop in what was substantially claimed as an unwarrantable and morally offensive statement made by him in a public sermon on the Sunday evening previous. After being assured by Mr. Cook that he did not refer to either of their shops they desired him to name the place to which he did refer. In his own parlor, in private conversation, he frankly, but — as the sequel proved — injudiciously, located the room in question. If we are correctly informed the information obtained from Mr. Cook was immediately conveyed to the proprietors of the shop, and they, as well as their employees, and the two manufacturers previously mentioned, came out with a card to the public, denying the charges of immorality in the statement to which we have several times referred. That much indignation should be felt by those on whom the blow had fallen is not at all surprising, and it is peculiarly unfortunate for all concerned that the blow should have fallen where it is said it was least deserved. The moral status of the employees in the shop in question is reputed to be specially good, and it seems incredible that a man professing to preach the gospel should wantonly assail the characters of innocent females who earn their bread by daily toil, and this hinges the subject on a question of veracity between Mr. Cook and a male employee in the establishment, the latter flatly denying that he ever used the language quoted by the former and laid at his (the operative's) door. Although Mr. Cook gave the language — at which special offence has been taken — as a quotation, and confidently asserts that the male employee told it to him, yet in view of the fact — we have no doubt that it is a fact — that the moral condition of the shop mentioned will compare favorably with any in the city, we presume that Mr. Cook will make the *amende honorable* to the female employees of the establisment. Unless he can show that the statement made upon heresay — as he claims — is true, he owes an apology to the ladies in the shop, and justice demands that he should make it.

On last Sunday evening the attendance at Music Hall was again very large, and after the preliminary exercises Mr. Cook read the following, which, at the earnest request of many citizens, we publish as read by Mr. Cook.

. . . .

This Mr. Cook followed by an excellent discourse upon "False and True Faith and Repentance," his text being, "What shall I do to be saved?"

During the closing exercises Mr. S. M. Bubier came upon the stage and began to address the audience in refutation of some of the remarks made by Mr. Cook in the previous discourse at the same place. A portion of the audience being about to leave the hall, Mr. Cook requested all to remain and listen to the gentleman. Mr. Bubier said in substance that he had no knowledge of any such place as had been alluded to by Mr. Cook, that he had lived in Lynn fifty-five years, had done business here much of that time, claimed to be better acquainted with shoe manufacturing systems than that gentleman, that man, who had resided here only seven months, and how long previous in an insane asylum he did not know. Mr. Cook made no reply, but the audience expressed their disapprobation of Mr. Bubier's remarks by hisses and cries of "Put him out." These demonstrations being repeated, Mr. Bubier was induced to leave the stage; after which the exercises were closed by singing a hymn.

In conclusion, it is but justice to Mr. Cook to say that, whatever his faults may have been, the conduct of the parties who interviewed him was hasty, ungenerous, and unworthy of men whose long life in the community added to their mercantile integrity had won the respect of large circles of acquaintances. Had they withheld the information obtained from Mr. Cook, the proprietors of the shop referred to might never have known who was meant, and respectable, hard-working women would not have suffered from an uncalled-for notoriety, and the public might have failed to guess where the arrow hit. That both sides have said things which they would be glad to recall we have no doubt; and with all due respect for Mr. Bubier, we have no excuse for his extraordinary conduct on Sunday evening last, even though it be, as claimed, that his motive was to protect the character of the working people of Lynn. Believing that the favorable representations made to us in regard to the characters of the females employed in the shop mentioned are true, we think the Rev. Mr. Cook's duty is clearly defined, and that the circumstances of the case imperatively demand an apology from him to the ladies referred to.

We would caution the working men against the effects of heated discussions on this subject, for discussions and differences in their ranks will lead to disastrous results. Remember the old motto:

"United, we stand,
Divided, we fall."

Little Giant, Feb. 4.

CARD TO THE PUBLIC. — While my sentiments upon the particular subject in discussion remain unchanged, I take this method of publicly expressing sincere regret for any improper action or unkind words into which I may have been betrayed by the excitement of the occasion last Sabbath evening.

Lynn, Feb. 2, 1871. S. M. BUBIER.

Semi-Weekly Reporter, Feb. 4.

LECTURE XI.

SUNDAY EVENING AT MUSIC HALL.

MUSIC Hall was once more filled to its utmost capacity last Sunday evening; the public anticipating that the subject which has agitated the community to such an extent within the past few weeks would again be reverted to by Rev. Mr. Cook, who was to conduct the meeting. The services were opened by fine singing from a chorus of one hundred voices, followed by reading of Scripture and prayer, after which Mr. Cook announced as a text the words found in John xv. 11: "These things have I spoken unto you, that my joy might remain in you, and that your joy might be full." The speaker prefaced his regular discourse with some remarks in allusion to the subject which has been so widely discussed of late, which, at our solicitation, have been kindly furnished us by him, and are as follows:

Henry IV. of France once said, and the remark has been gratefully remembered for two hundred years, that he hoped the day would come when every working man in his kingdom might have, as often as he pleased, a chicken for dinner. If there is a political ruler in the civilized part of the world who would not say this now, he is as much an exception to the drift of the age, and to the tendencies of history for the last two hundred years, as the Gulf Stream is an exception in the general drift of the Atlantic. But when Henry IV. said this, the speech was a singular one for a great ruler to make. Everything in history for two hundred years illustrates God's pity for the poor. De Tocqueville said that he regarded the progress of the democratic principle in governments as a providential fact — the result of a divine decree. It was universal. It was enduring. It was irresistible. All men and all events contributed to this progress. He found in it the sacred characters of a providential fact, and he stood in awe before it. But this progress is only one illustration of God's pity for the poor. The day has come for nearly the whole world that can be called civilized, when to say what Henry IV. said is to say nothing singular.

Tell working men and working women what you please, you cannot make them believe that he is not their friend who is the friend of their children. I have undertaken to maintain before you that the working men and working women of the manufacturing centres of New England have a right to ask of the capitalists and employers of those centres, that when the working people put their children into the workshops, where they must earn their support by daily manual toil from morning to night, the arrangements of those workrooms shall be such that the moral danger to their children shall be as little as possible. This is the outline of all I have said. *The working class have a right to ask this of the employing class.*

In putting before you this assertion, I do not regard myself as saying anything off the line of the speech of Henry IV. I do not consider myself as given to enthusiastic speech; for I hope there is something of caution in my nature, as well as of impetuosity. But if there is any line of thought on which I might throw away all my caution, and give full rein to impetuosity, it is the line of God's pity for the poor. And on precisely that line lies all I have said of the right of the working class in our manufacturing centres to ask of their employers such factory arrangements as shall make the moral temptations of their children as few as possible.

I wish to introduce before you a great and grave theme to-night, and had intended to refer to no other; but in order that I may introduce this other theme with the best effect, it is absolutely necessary for me to brush aside a little mist that lies between your eyes and mine. Fifteen or a score of lies yet walk about the streets; and, in order to get audience for a great theme of another kind here to-night, I must say something to prevent their tramplings being heard in Music Hall.

Six important points are to be kept in the foreground, so far as I have part in the discussion now agitating this city. I am anxious to put principles in the foreground and personalities in the background. These latter have had a ludicrous prominence in certain quarters for the last six days. Rufus Choate and Daniel Webster were once opposed in a legal case that turned on the size of certain wheels, and Webster had the wheels brought into court. When Choate had gone through an elaborate argument to prove that the wheels were not this or that, Webster's only reply was: "Gentlemen, there are the wheels." I wish to keep the great points prominent in the discussion which now has the ear of this city. The facts which I discuss lie close at hand. The wheels are in court.

1. *The lasting-board is not a sieve.* The door through which operatives in the factories of a great town like this are engaged is not a moral sifting-machine. The advertising board is put out at the door or window of a factory, asking for operatives for this class of work or that. In most factories, in the two great brisk seasons of work, whoever comes first that is skilful is likely to be taken. Orders must be filled. Teams must be made up. The operative here signs no regulations, as the operative at a Lawrence or a Lowell cotton-mill does. This is a most important peculiarity of a shoe town, as compared with a cotton town; and, on account of the two circumstances I am about to mention — the periodical lulls in the activity of the shoe factories and the large percentage of changeable operatives — it is difficult to introduce into the shoe factory system the admirable method of sifting operatives according to characters that has long been practiced in the cotton factory system, which does not have to meet the exigencies of such lulls and such a percentage of changeable operatives. A study of the subject at a distance might lead to the opinion that a shoe town and a cotton town are much alike; but the reverse is true. Each has a set of exigencies of its own; and mistakes on the whole subject of the shoe factory system arise from nothing oftener than from the false notion that a shoe town and a cotton town are parallels at most points. The sifting of character at the door of entrance of the workrooms of the factories of a city like this is not, and is not likely to be, by any means microscopic;

and this is my first reason for the separation of the sexes and the appointment of good moral men as overseers in the workrooms.

2. *In the shoe trade there are in each year two lulls, each of about a month's duration, and thousands of operatives are dismissed from work in these inactive seasons.* This makes the operative population of a city like this, of necessity, more floating than that of one given to the cotton trade. These lulls do not occur in the coal, the iron, the woollen, or the cotton trade. They are a peculiarity of the shoe trade; and the very best authorities I have consulted among the leading manufacturers tell me that, for the reasons I explained on a former occasion, these lulls are a necessity in that trade, and are likely to characterize it for scores of years to come. This is a most important peculiarity of a shoe town, as distinguished from the cotton town. *The floating population is the more floating on account of these lulls.* I have made no sweeping charges against the character of even this floating population. I said publicly, in the address which I gave two weeks ago, that I believed that if this population had homes, and was not floating, it would be as free from moral perils as any population of its size. We all know, and we all rejoice to know, that *it contains hundreds of most excellent people.* These very people, however, will be the last to accuse me of setting in too strong a light the perils of a floating population. Homelessness, I have said, is always a moral peril. And where, as here, a vast floating population congregates, the moral perils will always be found to be very great. He is pitiably ignorant of New England, and of this city, who does not know this. Now, it is out of this floating population, itself not sifted, that operatives are taken into the workrooms through a door that is itself not a sieve. And this is my second reason for asking for the separation of the sexes in the workrooms and the appointment of good moral men as overseers.

3. Thirty-three per cent of the operatives in your factories are changeable within the year, and you could not find out

their characters if you would. In view of these circumstances it is very far from extravagance to say that, in a city like this, the chances are extraordinarily great that bad men and bad women will occasionally be found in the workrooms. This would be a probable theory antecedent to experience; but this city, and other cities having similar interests, exhibit an experience about which I should convince you that I know nothing, if I did not say that the experience justifies the theory.

4. *Foul mouths in factories are so well known that the expression is almost a proverb.* I make no unguarded charges. There are numerous and most honorable exceptions, especially in the factories managed on the clean system; but you would think me ill acquainted with the most esssential parts of the subject I discuss, if I did not refer to what the best class of working men and working women speak of to me at every street corner. I need not expand this large consideration. It is a reason why the sexes should be separated in the workrooms. Unimportant as the circumstance may appear, it is not one of the least significant of the differences between a shoe factory and a cotton factory that in most of the rooms of the latter the machinery does, and in most of the rooms of the former does not, make noise enough to prevent conversation.

Utterly futile is it to compare, as I have heard a few white-handed critics carelessly do, the influences of the sexes on each other in these workrooms, filled at hap-hazard from a floating population through a door that is not a sieve, with their influences on each other in a high school. Woman's refining influence on man no one shall value more highly than I, but there is a proverb moderately well known among sensible men, to the effect that circumstances alter cases. (1) In a high school the immoral are excluded. (2) The sexes mingled there have homes of their own, and are under the restraint of social ties close at hand. (3) They usually are persons who have refined tastes. (4) They are constantly under the eye of a teacher.

5. Physicians in this city, long resident here, calm, scholarly, well-balanced men, who have opportunity to see what a minister rarely sees, and what an employer sees as rarely, solemnly testify, after making careful allowance for every explanatory circumstance, that the influence of the mingling of the sexes in the workrooms of the vast trade of this city has been detrimental to the floating population in a very high degree. I must ask any one who does not see reasons for the separation of the sexes, and the appointment of good moral men as overseers, as two measures of importance in the factory system of the largest trade of the United States, to study this subject most carefully, as I have endeavored to do, from the point of view of judicious and candid local physicians.

6. Since the population is more largely floating, and the percentage of changeable operatives so much greater in a shoe town than in a cotton town, you cannot do as much with boarding-houses in a shoe town as in a cotton town. I think four or five boarding-houses here, under as good moral management as that of the best in the cotton towns, would be excellent capital, and ought to be erected; but any system of corporation boarding-houses, such as exists in the cotton towns, is almost an impossibility here, owing to the fact that you have five or seven thousand people who are here or not here, according as your vast business is at its brisk seasons or at its lulls. *You cannot do as much with good boarding-houses in a shoe town as in a cotton town.* And therefore, since many of the moral measures that are commonly applied in manufacturing centres are inapplicable to shoe towns, those that can be applied become all the more important.

There are the wheels. Whoever does not keep fully in view these six points does not understand the complication of the problem of the moral condition of the shoe trade in a great city. My object here and now, as you notice, is only to keep before the public mind the important points of the topic, and to push personalities into the background. These are only very slight and inexhaustive hints of the reasons I have given for the separation of the sexes in the workrooms and the appointment of good moral men as overseers. I solemnly believe that if all the factories of this city were managed as some five or six among the largest of them already are, on the system of the separation of the sexes and the appointment of good moral men as overseers in the workrooms, more good would result to the city than any ten churches here can do in a year. (Applause.) I beg you, my friends, not to applaud — if you can help it. And I believe that the factories here that do not attend to these points, but are more or less careless of the moral effect of the arrangements of their workrooms, do more to injure the moral condition of the city than any ten churches can do in a year to improve it. (Applause.) I stand for the right of the working men and women in the largest trade in the United States, now organizing a factory system of more importance than the cotton factory system, as it is to cover a larger public interest, to ask of their employers such workroom management as shall make the moral perils of the working class, and of their children, as few and small as possible. These perils will be great enough in any case. The working people have a right to ask that they be reduced to the smallest amount possible. This is the substance of all I have said; and I shall take back my words when Henry IV. takes back his. (Applause.)

Two lies are in circulation that I must despatch, and then the way will be open to the other theme I wish to introduce.

1. It is said that I have attacked the ministry of this city. It was my fortune yesterday to walk to see the really great painting of the Battle of Gettysburg, now on exhibition in one of your halls, and to carry at my side an innocent opera-glass in a morocco case. If, in consequence, I am reported to have walked the streets in daylight with a revolver, I shall not be surprised, judging the future by the past. Patrick Henry said he had no means of judging the future but by the past. It is wholly inaccurate to say that I have attacked the ministers. Whoever heard me two weeks ago to-night will remember that I quoted Tennyson's phrase about the "lily-handed, snow-banded, dilettante priest," but did not quote the last word,

using "critic" instead of it; and I had in mind no minister. I said the theme was complicated, and that every minister must direct his own course of study. I would not have dared to speak on it myself without three months attention to it. It may very well be that, in the midst of multiplied activities, this or that public speaker, heartily interested in this theme, does not take it up because it cannot well be touched without a large attention to it. Speakers must choose their topics according to the course of their previous studies. The adaptations of men to this or that line of effort differ; and nothing is more juvenile or mischievous than that good men, taking different lines of effort according to their different adaptations, should therefore seem to themselves or to the public not to agree. All my relations with the ministry of this city are, and always have been, cordial in the extreme. I must say that I have myself seen service in battle at other points of the line of the enemy, and therefore sympathize with those leading the attack at those points, and have no fear that I shall be understood as underrating the importance of attack at any other point on account of attacking this. I confess I think the moral condition of this city a legitimate subject for its pulpit; and that the moral condition of the factories is a highly important part of that theme, and not the less important for being endlessly delicate and complicated. Those who labor in respect to this matter indirectly will not blame me for attacking it directly. The theme is not only complicated, but new; and, as the years pass, the responsibilities of the pulpit in regard to it will increase.

2. In regard to a matter which has been persistently used to turn against me the working men, I have nothing to say, except that it has been used in vain. It is in no sense other than that of a reiteration or more specific statement of what I have said or implied before, that I allude to the matter now, in order to make the air wholly clear for the theme which I wish to introduce.

(1) Before my lecture of January 22d, I said repeatedly in private that I had no charges to make against the *present*

character of a certain room, the place of which has been so unnecessarily mentioned in the public prints. Several gentlemen remember that they heard me say then, and I say now, that I think the *present* character of the room very good.

(2) I saw but six girls in the room, and one of these was pointed out to me as an exception to the rest. I vividly remember that one was pointed out as an exception. One of the six, I am told on excellent authority, has left the room since. To only four, therefore, of the eight whom I have been charged with accusing, would any remark of my own apply.

(3) What was said to me referred to a time some ten months or so ago. I never had the date accurately, except that the time referred to was some months before I visited the room. I never have said that what was said to me was *deserved.* I rejoice now, on different evidence, and on evidence of varied kinds, to believe that it was not deserved. I told, last week, the story of the three crows. Let me now tell that of the three *anys.* I said, last week: "If, by *any* misunderstanding, *any* injustice has been done *any* where, of course I shall be delighted to see it rectified." I could have chosen other illustrations.

(4) I do not consider myself responsible for the public reference to names of persons and places. I must regard, and I do regard, the publication of my exceedingly private statements, in a single conversation in my own parlor, a violation of implied confidence. Infinite mischief would come to society, if statements made in this way, in the confidence of parlor conversation, and touching reputation and character, are to be thrust in this style into the public prints. (Applause.)

Mr. Cook then proceeded to the regular discourse of the evening, the subject of which was, "The First Principles of a Scientific Christianity; or Christ's Gift of Fulness of Joy;" — an exhibition of the necessity of the new birth, and of the atonement. It was an able sermon, listened to with thorough attention, except slight restiveness near the doors, where stood a part of the audience. Towards the close of the meeting a person made some disturbance by referring to something that had been said concerning the fall of Paris; but he was taken out, and, as will be seen by the police record, dealt with in accordance with law. The next meeting will be held in the same place, next Sunday evening, when it is the intention of the committee having charge of the meetings not to open the doors until half-past six o'clock, as many had been standing an hour and a half before the address commenced, the hall being full at quarter past six. — *Semi-Weekly Reporter*, Feb. 8.

Mr. Cook's Discourse. — We wish to state that Mr. Cook's remarks last Sunday evening are not published at his request, but at ours. We learn that Mr. Cook wrote out his remarks immediately after the lecture, in order to prevent misapprehension. When they were applied for for the press, he revised them carefully and added a paragraph on the contrast between the workrooms of a factory and a high school. But our readers who were hearers will see that nine tenths of the language, and every word of the part applauded, are precisely that used in the extemporaneous remarks at Music Hall. — *Semi-Weekly Reporter*, Feb. 8.

The Rev. Mr. Cook and the Reform Question.

Mr. Editor: I am led to notice this question of reform in the shops of this city from the agitation lately made about them. As pastor here of the Catholic Church for the last twenty years and upwards, I think I ought not to be silent on a question affecting the reputation of this city, and of the people under my charge. I will venture to say that no person here knows so well the trials, the difficulties, and the troubles which are met with in families and in shops as I do. Besides, I have travelled considerably in this country, and have visited many cities and towns in England, Ireland and France, and I find here as much goodness, devotion and piety, and as few real evils, as I have found elsewhere; indeed, taking the population into consideration, I believe the evils here are less.

Now, so far as the Rev. Mr. Cook, who professes orthodoxy, advocates appropriate rooms or halls for girls and men to work apart, and that no communication should be held between each other during the hours of labor, and that competent women superintend the girls' department, and competent men the men's department, so far as practicable, I heartily concur with him; not because I condemn the present system, but that this separation would be more fitting, more conducive to good order, more beneficial for the public good, and more in harmony with Christian civilization. I am, then, in favor of a committee of one being appointed from each church, to meet, consult together, and attend to this matter. It is our duty as instructors, directors, and guides of the people to guard against abuses, and see that the divine truths which religion teaches may be made known, so as to promote the good conduct and the spiritual welfare of all.

But so far as the Rev. Mr. Cook used language to convey the impression of the immorality of girls and men working in shops and halls, I am opposed to him. The names of girls and men have been brought before the public, and he is said to have alluded to them as a vile and degraded class, corrupt and corrupters. This charge he has not denied, for he says he has nothing to change. He adds that some halls there are, "some seventy feet in length by twenty in width, where the moral condition is bad, and the moral danger of the arrangements great," — and thus the city is filled with these people, and filled with iniquity.

Now I say that this charge is a detraction and a calumny on the working girls and men of this city. The Rev. Mr. Cook makes it, and he sins grievously. Detraction is the unjust taking away of our neighbor's reputation, or the lessening of the good esteem in which he is held. This is done by giving publicity to faults hitherto secret. When the charge is false or the circumstances exaggerated it is a calumny. The Scriptures say, in Proverbs, chapter 24, verse 9, that the detractor or scorner is an abomination of men. And, in Ecclesiastes, chapter 10, verse 11, "The serpent will bite without enchantment, and a babbler is no better." Again, St. Paul, chapter 1, verses 30 and 32, says that backbiters, inventors of evil things, etc., are deserving of death. And again to the Corinthians, chapter 6, verse 10, "Neither drunkards nor revilers will inherit the kingdom of God."

Further testimony is unnecessary. Now the gravity of this sin is to be judged from the number of persons injured, from the damage they may suffer in their reputation, from the quality of the detractor, and from the number of persons who concur, or who may hereafter be led to concur, with him. Indeed, an evil report affecting the morality of a city is a diabolical sin, which tends to break up the peace of people and of families. In this case it is a sin against truth, charity, and justice. It displeases God, and tends to deceive men. It displeases God, who is the God of truth, and imitates the devil, who is the father of lies and of liars. It tends to deceive men by representing falsehood for truth.

I am told that Mr. S. M. Bubier, who has largely contributed to the prosperity of this city, being present on the occasion, manifested his disapprobation in unmeasured terms. He could not suffer the working-class to be abused, and the good name of of this city to be arraigned, without raising his voice against it. Had he left the hall, it might have been sufficient to show his disapprobation; but, fired with indignation, he wished to stand up boldly and manfully, and rebuke the foul detraction from the very spot it sprung. Indeed, hardened must be the heart — I had almost said soulless must be the man — that would attack poor, honest, industrious, and defenceless girls, who wend their way on foot, by the dawn of the morning, amid the cold blasts of winter, to these shops and halls, and when the labors of the day are over, tired and weary, at the shades of the evening they return to perhaps comfortless homes. Thus they toil to earn an honest livelihood for their parents, for their little brothers, and for themselves.

I have now presented my views on this agitated question, and entertain no unfriendly feelings to Rev. Mr. Cook. Indeed, I wish him every success in his efforts to improve the people of his charge. Respectfully yours,

PATRICK STRAIN,
Pastor of St. Mary's Church.

Semi-Weekly Reporter, Feb. 11.

MR. COOK AND HIS STATEMENTS.

MR. EDITOR: We desire, in as few words as practicable, and in behalf of the girls in our employ who were so falsely slandered by Mr. Cook, in the lecture recently delivered by him in Music Hall, to call upon Mr. Cook for an apology. Waiving all information which he claims to have received, we call him to an account for what he said in regard to the girls based upon his own judgment. Mr. Cook said in his lectures, in speaking of the shop, that when he entered, he picked out the girl who had resisted temptation; but that *all* the others were bad — so bad that no virtuous young man could remain in the room with them any length of time and remain virtuous.

Looking at the audience, he asked them how they supposed he was enabled to select the virtuous girl from the bad. He answered the question by remarking that after a woman had fallen, there were certain unmistakable signs by which the fall could be detected; that they had lost the flash and lustre of the eye, the bloom and rosy hue of the cheek, and in their place the haggard look is seen; and that was the condition of those he saw in this room, and that was how he knew.

Now, without having any desire to call in question Mr. Cook's vast experience in that particular direction, or that he may be eminently qualified to read the human face, still we think he must admit that in this instance his judgment was not infallible, and in consequence of his remarks being unqualifiedly false, and by their having been made, an act of great injustice was done, we call on him, as an act of simple justice on his part, to make a full and complete retraction of the remarks, in as public a manner as said remarks were made.

BERRY AND BEEDE.

Semi-Weekly Reporter, Feb. 15.

MR. EDITOR: In simple justice to a man who I sincerely believe has done and is now doing much good in this city, I desire to say a few calm words with regard to a card which appeared in the last issue of the *Reporter*, headed "Mr. Cook and his Statements."

In that card the reverend gentleman is accused of falsely slandering several girls, in a lecture recently delivered by him in Music Hall, and called upon publicly to apologize for and retract certain statements which twelve or fifteen hundred people in Lynn, who were present, well know he did not make. Having a clear recollection of what was said, and no possible motive to change the truth, I will, with your permission, state the facts in the case.

In the course of his remarks on what he has termed the foul system in factories, Mr. Cook said that he had visited a room, about twenty by seventy feet, in which were employed six girls, who he was informed by the overseer, were, with one exception, bad, and that at a previous time the character of the girls employed in that room was such that no young man could remain among them and retain his virtue; and while in the room he — indicating one of the girls — asked the overseer if she was not the exception of which he had spoken. The reply was, "Yes, she is." Mr. Cook said the others were coarse-looking. The speaker then went on to say, what I have heard him say before, that vice leaves its impress on the countenance, and that there are unmistakable signs which can be read, etc. In no part of that address was there any allusion made to any particular place or person, and the room in question was only designated by a guess at its size. No one of the large audience in Music Hall that evening could possibly gather from what was said where or from whom the speaker had his information.

But two gentlemen have since informed the public that they, in Mr. Cook's own house, and in private conversation, drew from him what under the circumstances can hardly be regarded as other than a confidential statement of how and where he got his facts. This, from what motives they best can tell, has been published, and thus they have erected a moral pillory on which is exhibited to public gaze the room and names of those employed therein. Is Mr. Cook responsible for that? True, his brush has been used, but it was taken without his leave, and he cannot be justly accountable for what smearing has been done. A HEARER.

Transcript, Feb. 18.

LECTURE V.

RELIGIOUS SERVICES AT MUSIC HALL.

THE severe storm of last Sunday prevented the usual attendance at Music Hall, although under the circumstances the congregation was satisfactory; indeed, in any ordinary church, it would have seemed large. Those who were present were nearly all males, in consequence of which the speaker, Rev. Joseph Cook, changed the tenor of his remarks so as to make them applicable to his hearers. Not being confined in the least to manuscript, and having a ready flow of thought and language, Mr. Cook is well calculated for any emergency of this kind. The opening service consisted of prayer, reading of the Scriptures, and singing by a choir of gentlemen, and was followed by a discourse, the subject of which was, "Christ's Human Nature the Ideal of Excellence for Young Men." Mr. Cook based his remarks upon a portion of the answer made by Christ to Pilate, in John xviii. 37: "I am a King." The sermon was commenced by statements in regard to interviews with Robert Tombs and Jefferson Davis, to illustrate the idea of treason. The young men of this generation might be considered as only a remnant, so many have fallen in the war. But those who remain should seek to sell their lives as dearly in defence of righteousness as did their brothers who died for the cause of the Union and Liberty. Treason to God consisted in intemperance, sensuality, gaming, and all the kindred sins to which man is addicted; and if they did battle on the side of Satan, the chief conspirator against the kingdom of holiness, the retribution of rebellion must be expected. Christ was the only King to whom allegiance is due; and the young men were urged in earnest language to enlist beneath the Christian banner. The audience were very attentive to the speaker's discourse, and many expressed themselves as being much gratified with the sentiments advanced. — *Semi-Weekly Reporter*, Feb. 15.

LECTURE VI.

SUNDAY EVENING SERVICES AT MUSIC HALL.

Music Hall was completely filled again last Sabbath evening to listen to Rev. Mr. Cook, a large number being gathered at the doors previous to their being opened waiting for an entrance. As the speaker stepped forward upon the platform he was greeted with applause, upon which he made the following remarks, previous to the opening of the regular service:

You will allow me, my friends, to say a word concerning applause. It used to be said in Rome that the tap of Caesar's finger was enough to awe a senate. Public sentiment is the modern Caesar. I am not insensible to the value of whatever is an indication of calm, intelligent, ultimate public sentiment. This is the fourth time that, on coming before this audience, I have been received in a manner not without its value as an indication of what public sentiment is concerning the general drift of what I have advanced. No one, not studying the matter through colored glasses, can consider me responsible for expressions which occur as I come before the audience. These expressions of yours have been made in spite of its being Sabbath evening, in spite of the place being used as a church, in spite of repeated requests on my part that applause be avoided, in spite of the critical character of much that I have felt it for the best interests of this city to say, and in spite of the extraordinary efforts of a few to turn the city against me; and they have now continued through four weeks. Lies against me have swarmed through this city for the last few weeks thicker than mosquitoes in a morass, or than leaves in Vallombrosa. You have not been misled by them. I consider this expression of public sentiment — all the circumstances taken into view — by no means without a certain value. Four weeks, I say, have now passed. It is an expression calm, intelligent, ultimate.

But, my friends, there is one institution of New England that has a fame to the ends of the earth, and which is in our keeping here to-night. It is a part of the glorious honor of New England, and we must all be moved by the feeling that some part of that honor is in our charge. I mean the New England Sabbath. The sacred character of the day is a motive too obvious and great and profound to need to be more than mentioned here. It should already fill all our thoughts. But I mention the historic character of the day only to add, what it is, however, hardly necessary to mention, that we must do nothing to bury any part of this glorious honor of New England in our father's graves.

I am aware that you are accustomed to use this beautiful and spacious hall for other purposes than those which now crowd it. But it is used now, for the time, as a church. And this audience is of good quality; and should not allow itself to be misunderstood by the public. I do not care to force my own preferences upon the audience; but I wish to say that I most seriously and earnestly prefer that we should avoid applause. At the very least, if Caesar shall tap his finger, I hope he will only tap; for that is enough for high significance, when it is Caesar. And if he shall not tap at all, I shall think that Caesar is all the greater Caesar.

The services of the evening were opened with prayer, and singing by a chorus of singers who kindly volunteered their assistance. The subject of the discourse was "Secret Sins as an Obstacle to Conversion," being founded upon the words found in 2 Samuel xii. 7: "Thou art the Man," and Ps. xix. 12: "Cleanse thou me from secret faults." The audience paid excellent attention, although many were forced to remain standing through the services. There will be no meeting in Music Hall next Sabbath evening, its occupancy by the members of Gen. Lander Post, for the purpose of a fair through the week, and up to a late hour Saturday night, preventing the return of the seats in time for the service without encroaching upon the Sabbath. — *Semi-Weekly Reporter*, Feb. 22.

The Labor Interests of Lynn.

The meeting at the North Congregational Church, in this city, last Thursday evening, was not as well attended as the importance of the subject demanded, but those present were evidently deeply interested in the matter. George W. Keene, Esq., delivered the opening address, and through his courtesy we are enabled to lay it before our readers, as follows:

Mr. President, and Ladies and Gentlemen: One week ago to-day the managers of this association very kindly and cordially invited me to make the opening debate this evening. I hesitated as to which subject would be the most useful to the cause of labor, and also the most suggestive for good. After deliberation it occurred to me that the "recent events," so-called, if taken up candidly from another stand-point, and if conducted free from personalities, would perhaps be the most productive of useful knowledge. I concluded to take that subject from the stand-point of a citizen and tax-payer interested in all the affairs of our city. With this explanation I will present you my views.

The consideration of the various statements recently made concerning the system adopted by the manufacturers of Lynn, and the liability of the present system tending to the moral degradation of the operators, has led me to examine somewhat into the facts, and review the situation of affairs in relation thereto. I do not assume to speak for any one but myself, and will present only my own observations, without the least unkindness to any person or persons who have been brought thus prominently before the public.

Assuming as a fair and honorable starting-point, and judging from what has been said and written, any fair-minded and intelligent person would come to the conclusion that the system of manufacturing shoes, as at present adopted in our city, was, in a moral point of view, *needlessly and recklessly bad;* and the manufacturers themselves, if not directly implicated in the matter, were regardless of its consequences. This view I think any stranger, knowing nothing of our people, would be likely to take from reading or hearing the public statements that have been made. While I do not intend to say that Lynn or its system of labor is perfect, or its inhabitants free from the temptations and vices incident to human nature, I do intend to say that — although we have been called upon to introduce an entire change in our mode of manufacturing, in a very short space of time, which has brought into our midst a great addition to our population, both male and female — we are, on the whole, and on the average, as clear in the particular matters charged as any of our neighboring manufacturing towns and cities. Yea, more. I think Lynn will stand, when fairly weighed in the scale of absolute truth, fully up to the standard of the best manufacturing city you may select, in point of morality and good citizenship. But even then there is enough to do to keep busy every benevolent and moral enterprise that can be started. There are liabilities to temptation on our right and on our left, and no benevolent and Christian person, desirous of doing good among us, will find himself long out of work.

Situation of Affairs in Lynn.

I have been intimately connected with the shoemaking business in Lynn, as workman and manufacturer, since 1826. I have watched carefully and interestedly the course of events through all the changes since that time, and have lived in

Lynn over a half century. It is not wise for us or prudent for the community to cover up or to exaggerate difficulties or defects that may appear in the situation or the moral status of any business in the city or State; but it should be examined fairly, in the light of common sense, with an honest purpose to apply, if found defective, such remedies as an enlightened Christian community shall sanction and commend. Every citizen must know that our city, from its connection with the manufacture of shoes, has become an important place, both in population and the value of its productions. They must also know that it differs in one particular from other manufacturing places; that its wealth and business is in the hands of individual enterprise, and not in large factory corporations. Our prosperous manufacturers are men from the common walks of life, who, after toiling and struggling with the vicissitudes of business, are themselves the controlling power, and personally interested in our success.

Separation of the Sexes.

It would not be surprising if some defects should appear in a system so suddenly introduced as the present mode of manufacturing shoes among us; for within the memory of most of us there were no females employed in our manufactories or in the shops of the workmen. The females found employment in their homes, binding and fitting the shoes, while the men made and prepared them. *Now* there are from three to five thousand women employed in the various factories in this city, in almost every department of industry connected with the business. They are found not only useful and efficient, but generally faithful and trustworthy. Shall we in this state of facts make the public statement, that the system is bad, because, working in shops with men, the women and the youth of our city will be degraded and given to immorality? Is this true? Do we find our public schools productive of immorality because the boys and girls go to the same schools? Do we find the family less pure and liable to corruption because composed of boys and girls together? Is it not the experience of all mankind, that what is true of the nature of boys and girls continues with us through life very much after the same order. Jean Paul Richter says: "To insure modesty I would advise the education of the sexes together, for two boys will preserve twelve girls, or two girls twelve boys, *innocent*, amid winks, jokes and improprieties, merely by that instinctive sense which is the forerunner of matured modesty. But I will guarantee nothing in a school where girls are alone together — still less where boys are alone." Do we find industry and labor productive of immoral conduct — or is it idleness and listlessness? I do not pretend to decide the exact fact, which is best — a separation of the sexes in the workshops or not; but this I think I can decide, that the chances for the result in the case of the five thousand women in our factories, engaged from morning till night each day in good, useful employment, or the selection of five thousand women out of any city in the country of equal population, without employment — however carefully they may be brought up — would be in favor of these Lynn shop girls making the best wives and mothers, and becoming the most useful citizens. Immoral conduct is not the peculiarity of Lynn, or in any manner the result of our system of business. It is as prolific in country villages, and every town and city, wherever it shall find nourishment in the hearts of the people. In those places where woman is held by man in the lowest esteem it is most destructive and degrading, and in those places where woman is elevated to the true dignity and majesty of her womanhood, there we find the purest morality and the highest virtue.

It is the solemn duty of all honest citizens to discountenance and discourage immorality, or the appearance of it, in society. But in the fair and honorable discussion of the subject, we should never forget the great primal law of our being, that we cannot separate man from his humanity; for in all the movements of society we see the liabilities and prerogatives of human nature go to make up the larger proportion of events in human progression, and are always, on the whole, onward and upward. What society most needs in the progress of human concerns is *faith* in human nature. It is not all bad, nor all vicious; and what every man and woman needs to sustain him or her through great trials and great temptations is faith that the community *believe in them*; and if you would render them more liable to fall into the abyss of shameful temptations, teach them to feel that the community doubt them.

Effect of Careless Expressions.

This, to my mind, is the desolating influence forced upon the public mind by the recent public discussion begun at Music Hall; it is a kind of an education which sermons and catechisms will scarcely undo. I think I can say, in justice, that every conscientious man in this community feels himself in some degree personally responsible for the public virtue; and from this cause comes the present conflict. Poisonous prejudices against classes are often instilled into the public mind by careless preaching or thoughtless expressions. For instance, let a person high in authority teach the people in Lynn that every person who makes a bad shoe loses the fire of his eye, and you will find every boy in the street looking into the eyes of his neighbor, and in time, if all his discoveries should be true, every workman would make bad shoes, and every eye lose its fire. And any statement of a similar nature, affecting the moral character of a class, is more disastrous to a community than even this.

Moral Aspect of the Question.

The girls employed in our factories are largely composed of Americans, and mostly from good, honest, Christian families. Is it fair or honorable, in any true sense, to hurl the imputation broadcast that these shop-women are immoral, and it would be dangerous to the morals of the young men to be in their company? Such a statement, repeated by an important member of our community, inflicts a wound upon society that will paralyze the efforts of the good in their kindly office of reform. Is it not rather the duty of every Christian minister, as well as every good citizen, to look fairly and honestly back to the cause? Every intelligent person ought to know that the human passions are the *most* important, yet the severest and most trying prerogatives belonging unto the race of man, and their correct discipline will make them the most exalted and noble principle that can render man honored and beloved of his race, and the beloved and honored in heaven. Too long, too long already, has this noble, God-given quality in human nature been wallowing in the mud and filth of stupid ignorance, writhing and struggling against ages of prejudice, hypocrisy, and sin, striving through all this shame and filth to make itself known and respected among men; for, when understood, it will be the greatest and highest principle that can produce peace, joy, and happiness. It is vested in our nature by our Creator for wise, holy, and exalted purposes. The violations of its high principles will surely return to us and our descendants in unmitigated sorrow. This truth is so solemn that no good citizen should ever dare refuse to utter or consider it; and I beg of you to remember that, while human passion is one of the elements which God has generously and liberally endowed us with, he has also given to us kingly and queenly authority over ourselves. To demoralize this high prerogative is to invite into our own *soul* a host of traitorous enemies, to destroy *all* that is beautiful and lovely within us. 'Tis the *idle* thought that corrupts us. 'Tis the *mind demoralized* that *lures* us into *error, vice, and sin.* 'Tis from within cometh the evil.

Call to mind that great and sublime truth uttered by Jesus, when he says: "He that *looketh* on a woman, to lust after her, hath committed adultery already in his heart." Can any person properly comprehend the tremendons meaning of these words, without considering the great amount of light and easy talk, the slanderous jesting about the virtue of the people, and the repeating of impure stories so habitually and carelessly, among all classes — rich and poor, upper and lower, white and black, male and female? Then ask ourselves if this is not cultivating and inviting lust, with all its issues, directly into the heart. To pick out the frailties and liabilities of poor human nature, and hold them up to contempt and ridicule, may be an easy task; but is it not catering to the lowest and meanest desires of the human soul? If we wish to do this, we shall find every human being in the universe frail enough for our purpose. We are all, every one, full of faults and imperfections, constantly stumbling, and constantly pleading at the Throne of Mercy for pardon. We know, also, that slanderous expressions and backbitings have a wonderfully prolific growth, which, if encouraged in any community, will sadly undermine, if not destroy, the efforts of the best, the truest, and purest laborers we have. *It is the very citadel of Satan*, his strongest breastwork and fortification, *the* battery from which his heaviest guns are discharged.

Without calling in question the motives of any one, I must give it as my solemn conviction, that the personal allusions that have been made, and which have found expression in our public journals, — whether denied or contested in point of fact, or not, — have caused a large increase of public scandal, greatly to the detriment of public morals. Moreover, the high position from whence it emanated has fixed in the minds of the people in our neighboring towns and cities a most foul aspersion against the moral character of our young and prosperous city, which it will take years of our best efforts to undo. And the worst of all is, it has sent into the midst of this community a whole series of scandals, giving rich food to nourish all the vile contemplations of sensuality, wherein dwelleth all the seeds of lust.

Undue Alarm as to Female Employment.

With this preface, I will give to you what I believe to be the condition of the present system of our manufactories, and the reason for the employment of female labor. Not that it is necessary for the defence of Lynn or her people, in respect to their morals or habits and customs, but because very erroneous opinions may be formed by very honest people in our own midst, as to the true situation of affairs among us; and because, if silent, it might be inferred, by equally honest people, that the whole budget of statements *were true*, and could not be fairly answered. With all kindly feeling and honorable regard for every good intention and true purpose of reform attempted to be instituted in the recent discussion concerning the morals of these workpeople, it does appear to me that they must have been misinformed, or, through alarm, must have misapprehended the facts in the case. The manufacturers of Lynn are certainly not the men to shrink from any moral responsibility that may publicly rest upon them, and, had they been consulted by any one with any sincere desire of doing good to their workpeople, would have found efficient aid in this most useful field of labor. I venture to assert that there is not a manufacturer among us who, if his attention was properly called to any defect, either in his building or his arrangements of labor, but would gladly avail himself of the hint, and, if in his judgment correct, would adopt it.

Shoemakers generally well educated.

The Lynn shoemakers — or, more properly, those brought up under the old system — are, generally speaking, very well and properly informed upon all subjects, both in national and state affairs, in private and public matters, and are not behind any other class in intelligence or understanding of the general observations of facts and fancies, and are not often known to be afraid to look facts squarely in the face. They have their own opinions upon all subjects, and are not afraid to utter them; and this is the class and style of men now controlling Lynn. The independence of the nature of their business encourages this liberal feeling in our midst. The first year the town of Lynn assumed to make an estimate of the valuation of her property was in 1830, when her whole taxable wealth was found to be a little rising $1,800,000! Our schoolhouses were small and inadequate, our churches poor and far between, and the dwellings of the people and shops of the manufacturers and workmen corresponded with the churches and schoolhouses. Let any intelligent citizen or stranger look over our city to-day, and he will tell you that in point of the arrangements of our streets, the convenient style and extent of our schools, the beauty of our private residences, the elegance of our churches and City

Hall, the stability and convenience of our manufactories, are equal to those in any of the manufacturing towns or cities in the State.

No Wealth from other Cities.

And all this has not come to us of itself; it has been accomplished by the intense labor and earnest efforts of our own citizens. We had no foreign help, no wealthy neighbors to pour into our laps their overflowing wealth. We had no inducements to offer them. We were only the makers of and dealers in shoes; and in earlier days this was considered not a very promising or fashionable business. Not only have this generation improved and adorned their own city, but, by upright and judicious conduct, have largely contributed to render the business of shoemaking not only honorable and respectable, but one of the most important branches of industry in the country. In the progress of this growth I could relate to you numerous incidents of thrilling interest connected with nearly every manufacturer in Lynn. One only will I relate, as it illustrates the character of our people. During the continuance of the World's Industrial Fair, at the Crystal Palace in New York, one of the Queen's Commissioners from England visited Lynn, in order to inform himself and his government how the workmen were managed in the shoemaking districts of this country. He was introduced to one of our manufacturers, and, after a strict examination into our mode of doing business, he expressed himself much surprised that we should trust our workmen with so much material to work up, without a stamp put upon the pieces by which to know them. Just at that moment one of his workmen came in with some work, and the manufacturer introduced him to the Commissioner, when an animated discussion arose between them upon subjects of state and national politics, religion, and trade. The Commissioner turned to the manufacturer, and said: "I don't wonder you trust your workmen, if this is a specimen, for he has beaten me on every tack."

Employment of Female Laborers.

The first attempt to introduce women's labor, to any extent, into our factories, was on the introduction of the sewing-machine, about the time of the adoption of our city form of government. Then the question of the moral propriety was fully considered and freely discussed by the manufacturers. Then our conveniences were unfavorable; there were no factories of sufficient size to properly accommodate them. Yet the attempt proved a success; and I think the universal experience of every manufacturer who thus early introduced the sewing-machines is, that the girls created a favorable influence in the factory. They earned better pay, and were more independent in their labors. And the moral result is practically this — that these girls thus early called together have become, in most cases, worthy members of society, and are now honored as wives and mothers among us. This I know to be the case in reference to the girls in two or three of the earlier manufactories. Since then, in the construction of every building for the purpose of manufacturing, every attention has been paid to the convenience and comfort of these operatives. The machines have been greatly improved, and also the mode of operating them so as to give more ease, as well as to guard the health of the laborer. Of course, there are many factories that are not large, yet still employ girls; but as fast as these manufacturers can enlarge their premises they do so, and always introduce the proper conveniences. The largest portion of the girls employed in Lynn are now employed in these departments, and I believe it to be the expressed opinion of a large majority of all the manufacturers that they are orderly and well-behaved, and entirely free from the loose and vicious conduct imputed to them. Every day their shops are visited by strangers, coming in at different times, and they are always found busily engaged, decorous, respectful, and orderly; and whoever visited the fair of the Grand Army, held at Music Hall last week, must have been profoundly impressed on seeing that great crowd of both men and women, each day and evening during the week, intermixed and intermingling throughout the entire building, consisting of thousands of our fellow-citizens of both sexes, and not a word, a look, or action observable by any one that could detract from the lady or the gentleman; and the ladies in these gatherings were largely composed of our Lynn shop-girls.

With the introduction of the McKay sewing-machine, a great change has been made in the mode of manufacturing shoes, by a systematic division of labor. With the favorable experience had in the employment of female help thus far, the manufacturer believed that many parts could be as well performed by women as by men; and wherever the circumstances are favorable, they invariably improve the condition of things around them. This system of making shoes has greatly consolidated our labor. Women are employed wherever their labor can be available, and in a large majority of factories, where they are at work with men, better order and decorum is manifest. The most important necessity felt by the manufacturer has been hitherto, is now, and always will be, the employment of the right person as superintendent. Experience only can prove what is the best mode of action in this comparatively new system of business. It would be the most disastrous thing that could befall the progress of woman's labor in our city, to have our manufacturers lose their faith in them. It is, therefore, the duty of every good citizen to do all in his power to elevate the true character of woman.

Business of Lynn self-made.

There is really less want of harmony between the employer and employed in Lynn than in any other place I am acquainted with, and there is really more kindly sympathy between the manufacturer and his workmen in our business than is usual in any other business. It is in the nature of business to bring manufacturer and workmen into harmony, because their interests are identical. There is in Lynn no aristocracy of wealth, or family. The leading men of Lynn shew prominently to the world that with us the poor of to-day will be the rich of to-morrow. This harmony of interests in Lynn is manifested by the meetings and friendly consultations of both parties last year, and the adoption of plans and arrangements mutually beneficial. There may be differences of opinion among us, but you will rarely find a business of so much importance as ours where the manufacturer feels that true interest in his workmen, and the workmen an equal interest in the success of the manufacturer; and the result has been (I am proud to say it) that Lynn today stands in the front rank with all the best cities in the Union, in point of business integrity and standing. And why should she not also in morality?

It is undoubtedly true that there is liability to an increase of moral impurity in all mixed communities, where the people know but little of each other. But every laborer, even in that field of reform, to be useful, must be patient, and by *wisdom* learn how to direct his efforts so as to benefit and improve that condition, rather than by haste to injure it. The efforts of the Woman's Mission for Christian Work, with many other associations among us, are silently doing noble work in this direction. Let any citizen go into our public schools, and there behold the thousands of bright and happy children gathered from every family in Lynn, see the good order, decorum and discipline, and reflect how cheerfully the public treasure is poured out, like water, to nourish and support them. And that treasure is furnished largely from the factory operations in our midst. Then ask yourself if the larger proportion of these beautiful children will not necessarily take their places side by side with the industry of this city, and become the operatives in these factories. Then what party is there in Lynn so cruel or thoughtless as to fasten the stigma of impurity upon these factories, which are and must be the hope of the city — the stay of the people — our only source of wealth? Rather let every citizen go forth clad in the rough habiliments of labor, ready to do every true and honest work in the cause of our children, in the cause of our city, and the cause of virtue and of God.

At the close of the address the meeting was announced to be open for discussion. Messrs. Henry Moore, Samuel Bancroft, Samuel Porter, Thos. Roberts, Nath'l Brown, C. J. Butler, J. Carruthers, Geo. W. Mudge, J. N. Buffum, and Rev. Mr. Cook took part in the debate; but our space will not allow of a report in detail. Mr. Cook announced that ex-Mayor Buffum would address the next meeting. He said that if he made any reference to what had been said it would be at Music Hall; but he would remark that "the representations that had been made of what he had said publicly were as full of error as an egg is of meat." — *Semi-Weekly Reporter*, March 4.

The following is a synopsis of the remarks of the different speakers who followed Mr. Keene at the Labor Reform Meeting, on Thursday evening last, which, owing to the crowded state of our columns, were omitted in our report of Saturday.

Mr. Henry Moore, principal of the Franklin-street grammar school, said that in the discussion of this question of Labor Reform — in fact, in the discussion of any question — *truth*, and not victory, or an unjust interest, ought to be the object; and keeping this in view, he did not see any occasion for anybody to get angry or excited. He thought it best to discuss this whole matter calmly and dispassionately. Many of the statements and arguments of the lecturer he should not dissent from, but he thought he had misapprehended the position of the gentleman who spoke in Music Hall. It was a very *easy*, but not a very *just* mode of argument, to place our antagonist in a *false* position and then batter down the false position. Such arguments might be very good for some purposes, but they utterly failed to touch the *real* position of our opponent. He did not think it just to accuse the gentleman who spoke in Music Hall of intentionally slandering Lynn, when he *opened* the discussion of this question with remarks *highly complimentary* to the enterprise, intelligence, and general good order and morality of the people of this city. There are moral evils and perils connected with the manufacturing system of Lynn, which every candid and intelligent person who has given the matter any thought, must admit. These evils may not be greater than exist in other large manufacturing centres, but yet they do exist, and what harm will come if we look these evils fairly in the face as honest men and Christians. The man who advocates sanitary provisions for a city is not generally considered an enemy to the health and and prosperity of that city. Mr. Moore thought the lecturer was unfortunate in his comparison of the mixing of the sexes in the shoe factories with the relation of boys and girls in our schools. For every one knows that the teacher is always present, that no talking is allowed, and in many schools not even whispering. For his part, he could not see any analogy in the two cases. He thought the lecturer had said some good things, but he had not touched the real positions of the gentleman who spoke at Music Hall.

Mr. Samuel A. Bancroft said that he agreed with the remarks of Mr. Moore. He said that he had worked in the shops of Lynn for twenty years, and he knew what he was talking about when he said that Mr. Cook had not told half the truth about the evils in these shops. He had followed the sea, and had travelled the country considerably in his lifetime, but he had never heard such vile and filthy language as he had heard uttered in the shoe shops of Lynn. If gentlemen wanted facts, he could find facts. What had been said at Music Hall concerning the manufactories was true. He had heard, in Mr. Keene's own factory, language the foulest that could come from a man's mouth. He said there was no use in trying to cover these things up, or to deny them, for everybody who knows anything about the matter knows that they do exist.

Mr. Samuel Porter said, that as he understood it to be in order to discuss this or former lectures, he wished to say a few words in opposition to one of the positions taken in a former lecture, in regard to the Crispins. The lecturer did not say Crispins, but he knew he meant them, when he talked about some of their positions not being democratic, about their having rotten timber in their vessel, which, if not removed, would sink it. Mr. Porter said that he would cite three illustrations to prove that their position was democratic. The illustrations cited were the tariff, life-insurance, and the church.

Mr. Thomas Roberts said that he thought it was a very poor plan for people to get angry every time anyone said anything which did not harmonize with their views. If, in the course of his remarks he should say anything which did not harmonize with the views of any one present, he hoped that they would not get angry about it. He said that he did not believe in the Crispin organization. He was opposed to a monopoly of muscle, just as much as to a monopoly of money. He opposed it, because he thought that in the end it would injure the working class more than any one else. He had contended that the strike of 1860 injured the working class, and benefited the manufacturer by enabling him to clean out all his old stock of goods. He said that he heard Mr. Cook's lecture in Music Hall, and he thought that he was talking for the best good of the laboring class, and they pretty generally so understood it.

Mr. Nathaniel Brown thought these meetings would be productive of good, if we only reached truth in the end. His remarks were not confined particularly to the address of the evening, but embraced various subjects, one of which was land monopolies, which he said he had always opposed.

Mr. Charles J. Butler remarked that a good deal had been said in these discussions about being born in Lynn. He had not that to boast of; but he trusted he was just as good a citizen nevertheless, and had the interests of the city as much at heart as any other resident. He had imbibed some pretty strong American ideas, and cited an incident to prove it which occurred during a recent visit "across the water." In company with several gentlemen they met another, whose outward form was the same as their own, but to whom his companions tipped their hats, in token of obeisance. Upon inquiry, he learned that it was Lord somebody, and that it was customary in that country to salute the gentry in the manner related; but, he said, he cherished too strong American sentiments to tip his hat to even a lord, or to any one else, on account of his wealth. He afterwards learned, too, that this lord was an oppressor of the poor, as was oftentimes the case. Mr. Butler said he did not believe in looking up to any man as superior, *merely* because he possesses wealth. He thought it important that the capitalist and laborer should understand each other better; and one important point to be ascertained in these discussions, when evils are found to exist, is, *how* to apply the remedy. He believed that, throughout the city, in nineteen out of twenty workrooms, both males and females would vote for a separation of the sexes; and the passage of Scripture referred to by the speaker this evening (Matt. v. 28) is one of the strongest arguments why they should be separated. This being so, let the females be placed by themselves, and thereby remove the temptation.

Mr. John Carruthers said that he regretted the lecturer had left the hall, as he wished to call his attention to some points in which he had done the speaker in Music Hall great injustice. He had had no intention of speaking, and he did not then intend to discuss the question of shoe factories, because of his own knowledge he knew but little about that question. But with regard to what had been said at Music Hall, he *did* feel competent to express an opinion, for he had attended all the meetings. These continued misrepresentations of what was there said were getting to be somewhat tiresome, and had been exploded over and over again. The position taken by the speaker at Music Hall had never been met. He had never made any attack upon Lynn, for he has said many times that he was proud of Lynn, old Lynn; and the evils in the factory system, pointed out by him, he was always particular to say, applied to what he called the "floating population." Every one who was present will remember the expression, "The lasting-board is not a sieve"; and in the fact that it is not a sieve, are to be found the dangers of the mixing of the sexes. He never said that men and women could not associate together without corrupting each other; but always under proper restrictions. If any injury has resulted to the reputation of Lynn abroad, Mr. Cook is not responsible, but the parties who paraded this matter in painted and perverted colors in the public prints. Let the responsibility for any injury done to the reputation of Lynn abroad rest where it belongs. Whatever may be our opinions with regard to the positions taken by the speaker at Music Hall, let us do him justice, and not discuss and criticise what he never said.

Mr. George W. Mudge thought it was a great mistake if any one supposed that happiness was an attendant upon wealth. . . .

Ex-Mayor Buffum, who was the last speaker, said he thought these meetings would be productive of good. They carried him back to the old Silsbee-street debates; but as it was late, and he had been invited to lecture at the next meeting, he would not then speak at length. — *Semi-Weekly Reporter*, March 8.

The License of the Pulpit.

Mr. Editor: The late Doctor Cooke, whose name and fame will be ever held in reverence by every true citizen of Lynn, has left on record this quaint comment, not altogether complimentary, perhaps, but nevertheless believed by him to be true, that, "if Satan has any degrading and filthy delusion to play off, to bring the human race into greater contempt, Lynn is the place which he is wont to choose for its birth." Had the great and good man foreseen that his own pulpit was destined to serve as the special birthplace, he would not so willingly have laid aside the armor in which he had so long and faithfully served; yet it seems to have been reserved for his direct successor to prove that the satire was not altogether undeserved, and that the occasion for it did not die with the genial critic who originated it.

Our community has been considerably excited, of late, and certainly not without good cause, by certain wholesale accusations publicly made by the present Rev. Mr. Cook, from what he regarded for the time as his own pulpit; and, although certain mild communications, both in support and in attempted mitigation of the offence, have from time to time appeared, it seems to me that the press of our city has hardly kept pace with public sentiment in the matter; remaining silent, doubtless, from a delicacy commendable in itself, and hoping that, through silence and neglect, the affair would die a natural death. But such a fate, however desirable, seems hardly possible in this case. The slander has found ready ears outside, and the rival manufacturing cities are only too glad to believe and spread the report. The press of these cities has repeated and commented upon it; and if, as we claim, the slander is baseless and false, they have a right to expect its refutation here, where it originated. Our manufacturers are daily in receipt of letters from their correspondents, really of regret and condolence over the depraved state of our society, taking the truth of it for granted, as either admitted or proved. Our former fellow-citizens, now located elsewhere, write to their friends in sorrow that their native city has so degenerated, and lament the change in our business that has brought with it the seemingly inevitable attendants, vice and immorality.

From all this, it is painfully evident that the fair fame of our city has suffered, and not inconsiderably, from this wanton attack upon the character of its citizens; and yet the reverend calumniator, though earnestly appealed to, wraps himself in his clerical dignity, and doggedly refuses "to bate a single word or take a letter back." Far be it from me to detract from the dignity and sanctity which New England education or prejudice has thrown around the sacred desk and the office of which it is the type and symbol; yet we deem it no heresy to regard those filling it as men of flesh and blood like ourselves, and with like passions, prejudices, and responsibilities too; and a slander, public or private, can claim no immunity from the laws of honor or of society because it issues from the pulpit.

As to the nature of the offence, as we choose to call it, of the reverend gentleman, no comment is needed, for it is already too well known to all of us; but a word as to the defence offered by his friends, since he himself does not condescend to unbend himself, either in defence or explanation, and which appears to be simply this that, so far as the slander was fastened directly upon individuals, it was a private conversation, and should not have been divulged; or, in other words, that only the *public* virtue and *general* reputation of our city were *publicly* attacked; *private* character and reputation having been allowed the privilege of a private and confidential execution. How this helps the matter, or mitigates the offence, I fail to see. That conversation has been divulged, and was first made public by his own friends, and has since been published in full, and in the public mind is inseparably connected with, and a part of, the general slander; and, more than that, the author still maintains it as true, and we must take him at his word.

All that is, however, immaterial here, for it is with the public and general slander that we have particularly to do. And to that point we submit, that it is a fair and logical deduction from the reverend gentleman's discourse, that among the operatives in our city, disorder and licentiousness constitute the rule, and good order, virtue, and chastity, the exception. Certainly we find no outward manifestations of this lamentable state of things. We defy the reverend gentleman or his supporters and friends, to point to a city of the business and population of Lynn, where fewer offences against the law are committed, or where good order is so distinctive a feature of all public occasions and places. A lady unattended can walk our streets, at any reasonable hour of the day or night, secure from harm or insult, and our police record will compare most favorably with that of any town or city in the Commonwealth. Such, we submit, is the character of our city, and we are each and all justly proud of it, and it certainly seems cruel and unjust that a reputation so fairly earned should be thus publicly assailed from individual caprice or ambition; and still worse, if the attack is the offspring of a diseased imagination, that it should receive moral support and character from the endorsement of the church, society, or community to which its author temporarily belongs.

It is not unreasonable to suppose that a person of ordinary prudence, before assailing the reputation of an individual or a community, would possess himself of facts or information in his justification. Yet here no facts are cited, and those whom the gentleman has quoted as his informants publicly deny their responsibility. But the reverend gentleman does not depend upon facts or proofs — a magical power of insight, vouchsafed to him alone, explains it all. He claims to possess a power of vision so wonderful that he can at will pierce the ordinarily impenetrable outside surface, and clearly discern secrets unrevealed to ordinary mortals. This power is intensified to infallibility at a certain season of the year, and, unfortunately for the victims, they were submitted to his scrutiny during " his month " — " the eye month," the wierd month of January. We have been educated in the belief that " the secret things belong unto the Lord our God "; but that was written long ago, and this is New England and the nineteenth century, and here, in the midst of us, is His vicegerent, with the divine attribute fully developed and in practical use, and each and all of us, with our wives and families, must pass at his option in review before this infallible judge, and be stamped and catalogued as to our fealty to virtue or vice, in accordance with his decree. " Whence hath this man this wisdom, and these mighty works ? " Where such a power exists, is it unreasonable that we should look for a commensurate responsibility; and have not the public whose honor is assailed, and the individuals whose characters are defamed, a right to demand where such responsibility rests ? If the author of the calumny is irresponsible, as we in charity believe, where else shall we look save to those who have publicly endorsed his course, and so given it its only claim even to public condemnation ? His church has already done so officially, and so have many of his over-zealous friends. Surely such zeal must be encouraging to a pastor, and is certainly commendable so long as the pastor is right; but it is simply intolerant bigotry when he is wrong.

Meanwhile, outraged public sentiment righteously demands of that pastor, and those of his people who uphold him, that they justify the slander by its proof, or retract it as publicly as it was uttered. LYNN.

Semi-Weekly Reporter, March 4.

Mr. Editor: — When we consider the various desultory attacks which have been made upon the position assumed by the Rev. Mr. Cook in his lecture in Music Hall, especially the attack made under the head of " The License of the Pulpit," we think it would be well, before wandering any further, to come back to the real and true statement of the case. The writer of the article referred to seems to have had a shot which he was very anxious to discharge; and in his anxiety he appears to have forgotten the fact that a gun too heavily loaded often kicks over the person who fires it. This writer quotes the language of the late Dr. Cooke.... What slander, sir, will compare with the language you yourself have quoted, and the author of which you yourself have commended? Neither the present Mr. Cook nor any one of his defenders is killed or wounded by that shot, certainly. If anybody is injured it must certainly be the man who fired it. He must have forgotten that his gun is of too small a calibre to shoot off Parsons Cooke without kicking.

This writer has a great deal to say about "*certain* wholesale accusations, publicly made by the present Rev. Mr. Cook." This is certainly indefinite enough. We challenge the author of this language to produce anything which the Rev. Joseph Cook has *himself* put in the public prints, which can by any fair construction be construed into a slander of Lynn. If *other parties* have given publicity to matters which have "found ready ears outside," and which the press of "rival manufacturing cities has repeated and commented upon," *let the responsibility rest on these parties, where it belongs.* No amount of sophistry can place the responsibility of this folly upon the Rev. Joseph Cook. If the author of this article cannot produce such evidence, let him retract *his* slander and misrepresentation.

This writer goes on to say further, "that it is a fair and logical deduction from the reverend gentleman's discourse, that among the operatives in our city disorder and licentiousness constitute the rule, and good order, virtue and chastity the exception." Now it would be difficult to conceive of a more deliberate misrepresentation than this. Every candid person who heard Mr. Cook's discourse, or who has read the published statements, knows that he has taken pains to say, over and over again, directly the reverse of this. But the author of this article seems to stand in mortal fear of what he terms "the reverend gentleman's magical power of insight." He seems to shrink from this "magical power of insight" as from some frightful apparition. We would say to our apprehensive friend, that innocent men and women are not apt to see apparitions, or fear any "magical power of insight." "The wicked flee when no man pursueth; but the righteous are bold as a lion." We wonder whether our apprehensive friend was ever alarmed at hearing a temperance lecturer say that he could tell an intemperate man by his countenance? or whether he never heard it said, before January, 1871, that the ravages of other vices than intemperance could be read in the countenances of men and women? I admit that it is rather an appalling fact to the vicious, yet none the less true. We would advise our friend, before he makes any more authoritative demands, to inform himself better in regard to facts. And we would inform the writer of that article that neither Mr. Cook nor his supporters are of that sort who cower beneath any man's lash. JUSTICE.

Semi-Weekly Reporter, March 11.

LECTURE VII.

SUNDAY EVENING SERVICES AT MUSIC HALL.

ANOTHER immense audience was present in Music Hall last Sabbath evening to listen to Rev. Mr. Cook, who had announced as a subject, "Poor Clothes, Rear Pews and Hard Work as Excuses for not attending Church; or the Value to the People of Sunday kept Holy." Such was the eagerness to secure an entrance, that, some time previous to the opening of the doors, enough people had gathered to fill the spacious hall; and every seat was occupied in a few minutes after admittance was allowed. The annoyance occasioned heretofore by passing up and down stairs after services had commenced, was obviated by closing the doors after every available space was occupied; and the congregation was very quiet and attentive. The meeting was opened with an invocation of divine blessing by Mr. Cook, followed by singing, and a fervent prayer from Rev. D. P. Noyes, of Boston, Secretary of the Massachusetts Committee of Home Evangelization. The regular discourse of the evening was then proceeded with; the words upon which it was based being found in Isaiah lviii. 13: "Call the Sabbath a delight."

Mr. Cook began by an allusion to Clark's Island, almost within view, in Massachusetts Bay, where the exploring expedition sent from the *Mayflower* rested, in spite of the necessity for labor and haste, on the Sabbath before the landing of the Pilgrims. Robert C. Winthrop, in his oration at Plymouth, at the recent two hundred and fiftieth anniversary of the landing, said that the Sabbath is of such importance that the most appropriate place for a monument to the Pilgrims is Clark's Island.

The following are simply the heads of Mr. Cook's address, without their development. The address was delivered wholly without notes.

1. The same difficulties between Labor and Capital which have arisen in Old England will arise in New England, as soon as our population is thick enough.

2. On account of the social and political position of the American working men, as compared with that of the European, Capital here, in adjusting its relations to Labor, cannot take the high and mighty method on the one hand, or the patronizing method on the other, both of which are so common in Europe.

3. It follows that, although the problem concerning the relations of Capital and Labor is the same in the New World as in the Old World, it must of necessity be treated in the New World, not in the Old World way, but in a New World way. The American industrial difficulties must have a peculiarly American treatment.

4. It will be found that the only

bridge which will carry a free Civilization safely over Barbarism and let its feet through at no point, is the Bible laid on the buttresses of the Sundays. The chasm between Capital and Labor can never be bridged in the United States, as it has been in England, by a kid glove. It can never be bridged in the United States, as it has been on the continent of Europe, by a bayonet. It can be bridged only by the Bible laid on the buttresses of the Sundays and the Common Schools. The nature of republican institutions is such that the chasm between Capital and Labor can be effectually bridged in the United States only by love, and not by force; only by intelligence and integrity in the opposing forces, and not by aristocratic prestige; that is, only by the Bible laid on the buttresses of the Sundays and of the Common Schools; and on neither the one nor the other of these two sets of buttresses taken by itself, though on the former alone much better than on the latter alone.

There exist, therefore, two sets of reasons why working men should make a right use of Sunday — the Industrial and the Religious.

I. *The Industrial Reasons.*

1. The right use of Sunday by working men, or public worship held by all classes as equals in one assembly, has an important tendency to diminish distance of feeling between rich and poor.

2. The right use of Sunday by working men has an important tendency to interest the largest, a well-educated, the most widely-heard, and the most conscientious class of public speakers, in the study of the wants of working men.

3. The right use of Sunday by working men has an important tendency to interest the largest, a well-educated, the most widely-heard, and the most conscientious class of public speakers, in the public discussion of the wants of working men.

4. On account of these three circumstances, the right use of Sunday by working men has an important tendency to make the church, which holds both rich and poor, such a means of allaying prejudice and securing mutual understanding, justice, and good-will, as to give it much the same relations to the different classes in society at large, as the Industrial Board of Arbitration or Conference has in particular cases of conflict.

5. Cheap Homes, Hours of Labor, Wages, Factory Reform, Temperance, all causes of the utmost importance to the industrial interests of working men, have among the masses of the members of the churches a vast amount of unexpressed sympathy, now largely wasted, and which a right use of the Sabbath by working men might do much, directly and indirectly, to turn into channels of the very highest value to those industrial interests.

6. The great principle of Co-operation, which is the hope of the cause of Labor, depends peculiarly for its success upon a high degree of self-control, integrity, and mutual confidence in moral character, between those who co-operate. This self-control and moral confidence have been proved, by the experience of associative effort in Europe and the United States, to be the cement without which the stones in the temple of Co-operation most assuredly cannot be laid; and the Sundays are

the trowels for the making and the laying of that cement.

7. It is the great interest of working men to preserve Sunday as a day of rest; but to neglect it as a day of worship is to undermine its authority as a day of rest.

These dead, backward, mossy churches are said by some to have no sympathy with working men. But they put Sunday on the statute-book, and keep it there, the only day of rest working men have. Undermine Sabbath as a day of worship, make it it only a legal holiday, and it cannot be preserved as a day of rest.

II. *The Religious Reasons.*

1. Sunday is God's day; the Sermon on the Mount re-institutes the moral spirit of the whole Decalogue; not one jot or title of the law, that is, of the Decalogue, is to fail; and, although Christianity transferred the Sabbath from the seventh to the first day of the week, the moral spirit of the fourth commandment is as fully re-instituted by the Sermon on the Mount as that of the seventh or eighth.

2. It is becoming more and more a proposition of science and scepticism itself, that the education of the moral and religious nature is the very highest of the interests of the individual man and of society at large. But this education is not furnished by schools, or adequately by literature. The best school civilization now presents for the education of what is highest in man is the right use of the church and of Sundays.

3. Children of parents who neglect the Sabbath are likely as a mass to go further than their parents in such neglect, falling behind, in moral and religious character, other children who have the benefit of the admirably equipped Sabbath Schools and of the whole right use of Sunday.

4. In a great city, the Sabbath is the worst day of dissipation.

5. Indispensable as legal aids to religion are, religion can never be enacted by law: the doors of the dram-shop, the gambling-house, and, indeed, of the poor-house, cannot be shut by reforms as to wages and hours of labor merely. The self-control induced by moral and religious education must supplement the effect of these reforms, or those doors cannot be shut.

6. An important moral and religious influence is exerted upon a population by its simply meeting once a week *in clean clothes and in sacred places and for public devotion*, whatever may be said of the poorness of sermons, although the average of sermons is as good as the average of Congressional speeches.

7. Sunday kept holy is a delight.

Poor clothes, rear pews and hard work do not keep the Catholic working men from attending church. Is it possible that American working men are more influenced by a necessity existing in these excuses than the Catholic? It is not enough to say that the Catholic church has free seats. There are other churches with free seats; but they are not full. It is not enough to say that a peculiar influence is exerted by priest over people. There are churches where personal affection between speaker and hearer amounts to an influence of a similar kind; but these are not full. If hard work is the excuse for not attending church, the ten thousand in this city who habitually neglect the Sabbath ought to fill the churches in the lulls of the trade of the city, which twice a year throw hundreds of workmen out of work. But the churches then are not full. Individual cases of hardship may exist in which these excuses have real force; but beneath the weight of the reasons that have been given for the right use of Sunday, they crack as an eggshell.

In regard to the topic now agitating this city, Mr. Cook said that five speeches were made at the Labor Meeting, Thursday, in reply to the address which opened that meeting, and which had been printed in full by the newspapers. The five speeches had not been reported. He hoped they would be. One of them was by a prominent teacher of the city; one by a gentleman who has been in the Massachusetts Legislature; and three by working men, whose daily life for years had given them such an acquaintance with the wants and sentiments of working men as only working men can have. The speeches had the marked applause, and plainly carried the assent of the overwhelming majority of the meeting, which consisted largely of working men. Of the nine ten-minute speeches made, not one supported the positions of the opening address. Mr. Cook said that it was unnecessary to remark that the speeches were utterly unprompted by himself. He sat in a rear seat; and so fully was the meeting carried by these speeches that he thought it necessary to say nothing at the Labor Meeting, except that, if he made any reference to the opening address, it would be at Music Hall, and that the representations which had been made of what he had said publicly were exceedingly inaccurate. When the five speeches should have been reported, it would be possible to make a more intelligible reference; and, accordingly, at the next meeting at Music Hall, he purposed to make some further allusion to the important topic now agitating this city.

The audience refrained from applauding when the speaker came upon the platform; but when he spoke of Scotch Sabbaths, Robert Burns, the working men, and at the close, marks of approval were given. — *Semi-Weekly Reporter*, March 8.

Music Hall Crowd. — The crowd on Sunday evenings assembled in front of Music Hall awaiting entrance to the religious service held there, has been so dense by the time appointed to open the doors, that the greatest difficulty has attended the efforts of the policemen in giving the uneasy throng admittance. On last Sunday evening, a Mrs. Brown, residing on Suffolk Street, was so seriously injured by being pressed against the door by the jam, that it became necessary to convey her to her home in a carriage. Would it not be advisable to open the doors at an earlier hour, and by thus doing prevent the recurrence of this disagreeable feature of a crowd? — *Little Giant*, March 11.

LECTURE VIII.

SUNDAY EVENING MEETINGS.

The eighth of the series of meetings under the auspices of the First Congregational Church of this city, was held at Music Hall on Sunday evening. The house was filled, a considerable number of persons standing, and a volunteer choir of singers occupied seats upon the platform. After the usual preliminaries, Rev. Mr. Cook proceeded to deliver, without notes, his regular discourse, the subject being, " The Necessity of the Atonement, an Inference from the Nature of Conscience and from the Unchangeableness of the Past," based upon words found in Romans, v. 11: " We also joy in God through our Lord Jesus Christ, by whom we have now received the atonement." It was a powerful argument in support of the doctrine of an expiation, in the future state, of sin committed while in the flesh. The universality of law admitted, and the important truth that no transgression, however slight, of either the physical or organic aws, is separated from the penalty, we can but conclude — reasoning from analogy — that the same is true of the moral law. As, therefore, in the natural course of things, no amount of good deeds can erase a single evil one, — the past being unchangeable, — the necessity of an atonement is plainly seen. If there is none, man, instead of being the happiest, is the most miserably cursed of all created beings. The positions of the preacher were enforced by apt illustrations, and he retained the marked attention of his hearers throughout.

At the close of the regular services, Rev. Mr. Cook proceeded to read from manuscript a statement he had prepared; a copy of which, revised by the author, we give below. — *Transcript*, March 18.

Previous to the dismissal of the audience, Mr. Cook made an allusion to the subject which has been under discussion for the past few weeks, in relation to the shoe factories of this city, which at our solicitation has been handed us for publication. — *Semi-Weekly Reporter*, March 15.

THE PROPER POINT OF VIEW IN THE DISCUSSION.

This audience is now hushed to the mood of prayer. Out of that mood let no man go, while, by taking the hands of this assembly, which represents so well every class in Lynn, I take the hand of the city that leads the largest trade of the United States, and ask you to meditate on a duty you owe to your own present and future, and to great public interests outside of your city to which your example has important relations. In Heaven's name, is a matter like this to be settled by petty spite and vindictive underhanded attack? That method of attack is being tried in this city. It is being tried by more parties than two or three, on more persons than two or three. I have never alluded, I do not now allude, to attack made on myself. But when others are assailed it is not so easy to be silent. It is a fit moment, in this hushed mood of prayer in which the audience now is, to appreciate the character of that style of attack. Out of that mood in which we now are, no man is fit to give advice. That mood is lifted high above the fear of man, above jealousies, above all childish and subterranean vindictiveness. That mood is one of equal tenderness and courage; of fairness and fearlessness. It is a spirit that will cut off the right hand sooner than commit injustice, but which laughs at the shaking of the spear.

It is not merely on local interests, but on interests vastly greater than local, that I have occasionally asked this audience to meditate.

1. The trade of which this city is a representative is larger than any other single branch of manufactures in the United States.

2. This trade is now in a state of transition, and is adopting a new system of labor which will affect the interests of a greater number than is reached by any other single branch of manufactures in the Union.

3. This city leads that trade; and the fashions it shall set for the new system are of importance far beyond the limits of your local interests.

My point of view has always been that which these propositions indicate. I came to this city refusing urgent invitation to stay here longer than a year. I have had for years on my list of plans the spending of a year or so abroad, for travel and study; and my present hope is to execute the plan within a few months. It is amusing to notice how the love of notoriety has been assigned to me as a motive. I have no interests that would lead me to plead for an audience in Lynn. The public understands that I speak disinterestedly, as well as dispassionately. My interest in this theme arises from its almost national importance.

DENIAL OF CHARGES.

If any one thinks that I should not now be the most unpopular man in this city, or that weeks ago I should not have been hissed off this platform, instead of being obliged to repress applause here night after night, if I had made the wholsale accusations against the character of this city which I have been so childishly accused of having made; or if what I said here in January of your vast trade had not been substantially true, let such an one go his way, light of heart. He must have a deep mind.

POSITION OF LOCAL PUBLIC SENTIMENT.

On all hands the evidence exists that the working men and working women, as a mass, are with me with a surprising degree of unanimity, both in respect to the matters of fact alleged, and especially in regard to the two remedial measures which were suggested. But they are so in spite of extraordinary efforts to turn them against me. They cry out for the clean system. I say again, as I said in January, God hear their cry! That this is the attitude of the best class of working men and working women in this city, is beyond all doubt. By the clean system, I mean the separation of the sexes in the workrooms, and care in the appointment of moral men as overseers. By the foul system, I mean the mingling of the sexes in the workrooms, and carelessness as to the moral character of overseers. My main assertion, as you know, is that the moral perils of the former system are less than those of the latter; and, until some one has disproved this assertion, understood in the sense of those definitions which I gave at the outset, no one has come within beat of drum of turning any central position of mine. But this is the position independently assumed by the overwhelming mass of the fifteen thousand working men and women now in this city; and who have, in an important degree, in their charge, the interests of the twenty, thirty, and fifty thousand that are yet to constitute your operative population. In this attitude of those the most interested and the best informed, I find an argument which, as the example of five or six of the leading factories here already proves, no thoughtful man, whose care for the interests of labor is not a pretence, will resist; and which those whose care for those interests is a pretence will in the end find irresistible, if working men and working women are but true to themselves in demanding, in every judicious way, that which they have such unimpeachable right to ask. I beg you to notice that I do not rely too much on such evidences as those which appeared at the Labor Meeting the other evening, when, after an opening address criticising some of my positions, five speakers immediately rose, wholly without prompting from me, to reply. One of the five, whose daily life for years has given him personal acquaintance, of an intimate kind, with that of which he spoke, gave it as his opinion that, should a vote be taken in the workrooms of the city, in nineteen out of twenty both males and females would decide to work in separate rooms. Another speaker said severer things on foul mouths in factories than I have ever said, though not severer than I have heard from other sources an indefinite number of times. That all mouths in factories are foul is not what I have asserted; but that, under present arrangements, a few foul mouths can corrupt a whole room. As to that danger, the assertion has come to me from working men more than a hundred times within a month, that I have not told half the truth. A third speaker held that what had been said at Music Hall was for the good of the working classes, and that they pretty generally so understood it. These speeches had the marked applause of the meeting, which consisted largely of working men. But they were only one, and comparatively a minor indication. I might hear of several meetings of an opposite kind and not have evidence enough to counteract the proof which exists elsewhere. I rely, in ascertaining the sentiment of working men, on wide and careful conversation with them, and on those who have had wide and careful conversation with them; but more than all on the tendency of many different kinds of evidence from differently prejudiced sources to the same point, and especially on the continuance of all this through time enough for opinion to become calm and intelligent.

Outside of the fifteen thousand directly or indirectly engaged in your great trade, public sentiment gives a powerful support to the general drift of what has been advanced here, as the frequent necessity of repressing applause in this hall on Sabbath evenings, and this for six weeks in succession, has only very partially indicated. Written expressions from the most respectable sources come to me, in which the writers take the position that the measures which have been recommended have the sympathy and support of good men here, irrespective of age or occupation; and thousands of verbal expressions of the same unqualified kind have reached me. I know how to make deductions in all expressions in my favor, and notice whatever occurs of an opposite character; but, if the deductions were half of the whole, enough would remain to prove all I am asserting of the general tone of public sentiment. One of the pastors of the city volunteered to me the opinion that his church endorses me as fully as my own, in its official and public resolution. It is not unimportant to notice that the Catholic pulpit of Lynn has put itself in print fully on the side of the two measures which have been suggested, although, on account of inaccurate information as to what I said publicly, denouncing me for injuring the reputation of the city. In view of the extraordinary misrepresentations which have abounded, this tone of public sentiment has the higher significance. Persistent efforts have been made to turn the city against me. But the city persistently refuses to be turned. Six weeks have now passed. There has been time for public sentiment to become calm and intelligent.

I am honestly proud of this city for its present attitude. From Boston to Chicago, Lynn has been praised in the religious papers, and in some of the secular, for that attitude. I must solemnly beg leave to inform the group of capitalists and manufacturers and the inextensive circle of their dependents, who constitute the only offended parties here, that there are twenty-seven thousand persons in Lynn who are not capitalists or manufacturers. To-day Lynn stands as a queen, clothed in the prosperity of her last twenty years, her schools and churches and libraries glittering among her jewels, and she puts one hand on the great factories here that are careful of the moral effect of their arrangements, and says: Give me more of these; and she puts the finger of the other hand in scorn upon the factories that are careless of the moral effect of their arrangements, and says: Give me fewer of these!

WIDE CONTRASTS BETWEEN A SHOE TOWN AND A COTTON TOWN.

Six circumstances of great importance distinguish a shoe town, like Lynn, from a cotton town, like Lawrence or Lowell. (1.) The shoe trade has, and the cotton trade has not, two periods of comparative inactivity each year, and thousands of operatives in those periods are thrown out of work. The best kind of shoes, and especially ladies' shoes, to the production of which this city devotes the largest part of its industry, change their fashions with the seasons. Cotton does not change its fashions. You may safely accumulate cotton cloth. But it will not do to accumulate a stock of outgrown fashions in shoes. The new fashions cannot be foreseen for months in advance. And besides this, the new machinery invented for the processes of the shoe trade makes production so rapid that many manufacturers assign over-production as the chief cause of the lulls At any rate, when the market is full, buyers have sellers at their mercy. The law of the trade is not to accumulate shoes, but to fill orders for shoes. At certain periods of the year, orders for shoes pour in and a brisk period of work ensues. The orders grow fewer, and a lull ensues. The orders pour in again when the new fashions have been determined, and brisk work follows once more. The orders wane and work wanes. For all these reasons it is to be expected that, while production continues as rapid as it now is, and by the invention of new machinery it is growing more rapid every day, lulls will characterize the shoe trade fully enough to distinguish that trade broadly from the coal, the iron, or the cotton, which produce articles in the very nature of which there does not inhere, as there does inhere in the very nature of the article produced by the shoe trade, a constant susceptibility to change of fashion. Various expedients can, and I hope will, be used to shorten these lulls; but all of the many manufacturers with whom I have conversed consider lulls an outgrowth of necessity in the structure of the trade, and working men know well that to counteract their effect is one of the problems of their industrial life. (2.) The percentage of operatives changeable within the year is much greater in a shoe town than in a cotton town, and is estimated in this city to be thirty-three per cent of the whole number. (3.) Largely on account of these two circumstances, the floating population of a shoe town is likely to be larger than that of a cotton town. It is very evident, also, that the floating population is likely to be not only larger, but the more floating, on account of these lulls and the large percentage of changeable operatives. (4.) I have before pointed out how the three circumstances already mentioned make it difficult to introduce into the shoe factory system the admirable arrangements for sifting operatives, according to character, which have been carefully studied and elaborately applied for a hundred years in the cotton factory system, and how this partial inability of the shoe trade to make the doors of entrance to its workrooms a sieve, is an irresistible reason why the sexes should be separated and moral men appointed as overseers in those workrooms, especially as the rooms must be filled largely from a floating population, larger and more floating than that to which a cotton town is likely to be called to adjust its superior appliances. *Everything cannot be done in a shoe town that can be done in a cotton town; but there is more need for something to be done; and, therefore, what can be done is the more important; and what can be done is to separate the sexes in the workrooms, and to take the utmost care in the appointment of moral men as overseers.* (5.) The most of the rooms of a shoe factory differ from the most of those in a cotton factory, in that in the former the machinery does not, and in the latter does, make noise enough to prevent free conversation between operatives; and, unimportant as this difference may appear, it is in itself, and especially in connection with the four circumstances already mentioned, by no means one of the least significant of the wide contrasts in the problems to which the cotton factory system and the shoe factory system are to be adjusted. In the cotton towns the greatest pains has been taken with the excellent corporation boarding-houses. Each boarding-house of this class accommodates either men or women exclusively. But I see no reason why the sexes should be separated in the workrooms under the control of the cotton factory system. (6.) The large factories here are individual enterprises, and not the property of corporations as in Lawrence or Lowell. The consequence is that if the leader of the business in any factory is disposed to be careful as to his arrangements, he can have his own way easily, for he has only himself to consult; and if he is disposed to be careless he can have his own way, for the same reason; and in the latter case can have it more easily than if in a corporation, competing, as every great corporation at Lawrence or Lowell does, with some other great corporation, and this in regard to the moral as well as the material welfare of the operative population.

EXIGENCIES LIKELY TO ARISE IN THE FUTURE OF MANUFACTURING CENTRES IN NEW ENGLAND.

So much on the contrast between a shoe town and a cotton town. Already this city has a population of five or seven thousand who are here or not here according as business is at its brisk period or at its lulls. How large will that population be in ten years? How large in twenty? I am in Lynn but for a moment; but I profess to care enough for it to keep fifty and a hundred years of its future in view, and to put at hazard any popularity I may or might have in this city, by asking you to meet, as men, the complicated problems of your vast industry, and not set a careless precedent for that crowded future. If the population of this city increases for the next twenty years as it has for the last twenty, you will have, before the end of that time, or soon thereafter, a floating population of ten or fifteen thousand.

Now, is there a man — *who* is the man and *where* is the man — who will say that you can have a tide of ten or fifteen thousand people swirling in and out of a city like this and no moral perils arise, no sediment be stirred, no grave responsibilities laid upon those whose business is the floodgate through which these tides must mingle with the other tides of the population?

At the best, the filter that you can provide for the tides will be ineffective enough; but to say that there is need of no filter; that you may safely take the chances of the present factory arrangements being continued here, is to say — what time will disprove. If the present careless factory arrangements are continued fifty years, you will have a city full of moral ulcers. Lazarus will lie at the gate of Dives in this city, and he will be full of sores. I throw my whole weight into the scale against the continuance of these careless arrangements. *I know that the American Lazarus may to-morrow, or in the next generation, become a Dives, as the European may not; but, in spite of American institutions, the day is coming, unless factory life is studied and adjusted most carefully, when here and throughout New England, of which the whole Atlantic slope is a factory, Lazarus will lie at the gate of Dives.*

DISTINCTION BETWEEN OLD LYNN AND FLOATING LYNN.

Just here the importance of the distinction I have drawn between Old Lynn and Floating Lynn can be seen. By Old Lynn, I do not mean Lynn as it was; and by Floating Lynn, Lynn as it is, as some have taken for granted who have wanted a pretext for the assertion that I have slandered the whole city. By Floating Lynn, I mean the population that is here or not here, as business is brisk or the opposite. By Old Lynn, I mean fixed Lynn, resident Lynn. In criticising the floating population, I do not criticise the entire mass of those employed in the factories. Thousands of those have homes here, and the homes are a great credit to the city. I have repeatedly eulogized in public the cottages and gardens which exhibit the excellent conditions of the larger part of the resident manufacturing population. I believe those conditions equal to those of any resident manufacturing population in New England, in any city of the size of this. Neither have I brought sweeping charges against the floating population itself. There are in it, as I have repeatedly said, hundreds of most excellent and intelligent people. But a floating population in Lynn, I must say, in the name of all common candor, is not so far different from a floating population elsewhere as to be exempt from the moral perils of homelessness, which no sensible man is likely to deny, and which are illustrated the world over. I have called attention to the circumstance that in those points of New England where vast floating populations congregate — and this city is one of those points — the moral perils will always be found to be great.

In the system on which your industry was conducted for a hundred years, men and women worked separately. The small shoe shops, each containing ten or fifteen men, had one branch of work; and the females, at home, another. Take the old shoe shops: the talk in them cultivated a kind of brisk intelligence, of which every shoe town shows traces ; but I have never heard even those old shoe shops commended, on the whole and on the average, as the best places to bring up a boy in ! Suppose the old shoe shops had been emptied into the workrooms of the women. Your population was nearly all resident then. My claim is that the danger of the arrangements now existing in Lynn is twice as great as the danger that would have arisen in the case supposed, your population is now so much more largely floating.

REPLY TO MR. KEENE'S ADDRESS.

In regard to the opening address delivered at the recent Labor meeting in this city, it is hardly necessary that I should say anything after the five speeches which were immediately volunteered in reply to it.

(1.) The address is courteous. I rejoice to recognize its courtesy.

(2.) It does not take ground decisively against the separation of the sexes. It says: "I do not pretend to decide the exact fact, which is best, a separation of the sexes in the workshops, or not."

If I say a frank word or so as to the argument of the address, I must not be understood as personal, for I am speaking of the argument, and not of its author.

1. As an argument, the address is up to its knees in what I call the high-school swamp. (Applause.) Five circumstances distinguish the relations of sexes mingled in a high school from those relations in the workroom of a factory. (1.) The immoral are excluded from the high school. (2.) The sexes there mingled have homes of their own and the restraint of social ties close at hand. (3.) Usually they are persons of refined tastes, or whose most important object is to acquire those tastes. (4.) They are constantly under the eye of a teacher. (5.) Religious exercises open or close the school. Now, when this address, written by the President of the Board of Trade, claims that the present arrangements of the factory system here have been carefully studied, and then gives as a specimen of that carefulness the overlooking of distinctions as wide and clear as these; and says that, because Jean Paul Frederick Richter — I have a high respect for Richter's authority — remarks that the sexes ought to be educated together, the inference is that they should work together in rooms filled at hap-hazard from a floating population through a door that is not a seive, it is not too much to say that the address is up to its knees in the high-school swamp. A not graceful position for the genial President of the Board of Trade!

2. The address is self-contradictory in its claims. It asserts and insinuates that Lynn, in respect to "all the matters charged," is equal to any manufacturing town in New England. It then puts forward the plea that the system here is very young, and that its imperfections must be excused on that ground. These two claims cannot be made in the same breath. It would be singular indeed if the shoe factory system, which has begun within ten years, were already as perfect as the cotton factory system, which has been carefully studied for two hundred. The "matters charged" are that the shoe factory system here exhibits carelessness on important points. That the system of Lynn is young is the one strong point of excuse for it. But it is getting late to call it young. Property that will not change its shape for fifty years is being invested in it here every day.

3. It is claimed in this address that as fast as the great firms here are able they make "all proper" arrangements for their operatives. I simply deny this. Many of the great firms do. But the poet Tennyson praises Prince Albert for devoting his time to studying the models of lodging-houses. The most delicate-minded of the English poets praises the noblest of the recent English princes for studying the arrangements for cottage workrooms, kitchens, and even sewers. Would that some Prince Albert and Poet Tennyson might criticise certain very wealthy firms here, who have incomes of a hundred thousand a year, and yet make only the most wretched and inadequate provisions on points Prince Albert did not disdain to study.

4. The address fails to meet the decisive points of the case. I do not call the argument evasive, for I believe it honest as well as courteous. It fails to meet the points that the advertising board, through which operatives are engaged, is not a sieve; that the percentage of changeable operatives is high; that the floating population of a shoe town is likely to be large; that foul mouths in factories are well understood to be a part of the problem; that physicians of this city solemnly testify that important evils result from the present arrangements;

and that corporation boarding-houses cannot be in a shoe town what they are in a cotton town.

When the hunter on the Amazon sees a certain shadow on the ground at midday, he knows that the Amazonian vulture hangs above in the air. I have not seen the vulture. I am much mistaken if I have not seen his shadow. Working men have said to me that they do not dare discuss this subject for fear their wages will be reduced. I was told by one of the best representatives of their interests that they feared to agitate this theme because of apprehension that the Board of Arbitration or Conference, between employers and employed, so sensibly organized here, might be disturbed in its operations. A gentleman of the highest intelligence and culture told me he prepared two articles for the press, in support of the improvements suggested, and one of them he read to me; but keeping in mind the treatment of certain parties in the past, he decided not to publish the articles, and was advised by his friends not to publish; for fear of unscrupulous attack by petty business annoyances from rich men. A most intelligent lady told me that she had prepared an article in support of the interests of the working class in this discussion, and was induced by her husband not to print it, solely because of the fear of unscrupulous attack by petty business annoyances. I have in mind a score of unimpeachable illustrations. More than twenty times, and I think more than fifty, I have crossed this shadow; and more than six times I have crossed it within my own parish. The fear of attack from rich men! This I call the shadow of the Amazonian vulture. I hope it is only a cloud of exaggeration which casts it. But when one sees the shadow twenty or fifty times, that is enough to found an inference on, stout enough to cause a man at least to look up. (Applause). — *Semi-Weekly Reporter*, March 15; *Transcript*, March 18.

LECTURE IX.

SUNDAY EVENING SERVICES AT MUSIC HALL.

Music Hall was again filled on Sunday evening last, nearly every seat being occupied; and many persons remained standing during the services. The audience was very attentive; and several good points were made by the speaker, Rev. Joseph Cook, who received applause, although a request had been made to the contrary by Mr. C. His subject was "Christ in the Family, and in all the Relations of the Social to the Religious Life; or, Cheap Homes and Boarding-Houses in Lynn." The text was chosen from James ii. 15, 16; the following words being selected: "Those things which are needful for the body." The discourse was an able and important one. The synopsis which we give below embraces a plan of the whole, without its development:

1. It is a remarkable fact that the growth of the average population of the cities of the United States between 1850 and 1860 was more than twice as great as that of the population of the whole country. In the census of 1860 a list of one hundred and twenty-six cities is given, scattered through the whole territory of the Union, and the average increase of their population between 1850 and 1860 was 78.62 per cent, while that of the whole population of the United States during the same period was only 35.59 per cent. In this period, the increase of the population of New York was 56 per cent; of Philadelphia, 60; of Worcester, 46; of Hartford, 115; of New Haven, 93; of Buffalo, 91; of Nashville, 62; of New Orleans, 44; of St. Louis, 106; of Chicago, 264.

2. The growth of the population of the cities of Europe is much more rapid than that of the whole of any single nation of Europe. From 1832 to 1869 the increase of the population of London was 98 per cent.; of Constantinople, 50; of Paris, 118; of Vienna, 107; of Moscow, 50; of Berlin, 220; of Manchester, 49; of Liverpool, 174; of Madrid, 105. Liverpool has increased as rapidly as Boston, Berlin more rapidly than New York.

3. It is evident from the two preceding propositions that there is at present a tendency of population, here and abroad, to mass itself in cities; and, it will be found on examination, that this tendency has been exhibiting itself in greater and greater force for the last fifty years.

4. The growth of means of intercommunication has been a chief feature of the last fifty years. The Atlantic Cables, the Pacific Railway, the Suez Canal, with the infinite multiplication of subsidiary railways and telegraphs, are intro-

ducing a new set of circumstances into civilization.

5. It will be found that the growth and influence of great cities increase with every increase of the means of intercommunication between States; but the increase of such means of intercommunication, as has just been shown, is very nearly the most characteristic feature of the present age. The current of population is governed by the law of supply and demand. The demand is for population at centres, since these furnish opportunities to industry and capital not found elsewhere. But the creation of such centres, easily intercommunicating, is a permanent result of the growth of the means of intercommunication.

6. The problem of the perishing classes in our cities increases in importance with every increase of this tendency of population to mass itself in centres. Every addition to the growth and influence of great cities adds importance and amplification to that problem.

7. Frontier life is essentially a struggle for food, shelter, and raiment. We honor that life on the slopes of the Rocky Mountains. It has an historical glory in the story of the first winter of the Pilgrims at Plymouth, where Rose Standish and half of those who came over in the *Mayflower* laid down in the graves on the slope of the hill across the Bay yonder, in a struggle simply for food, shelter, and raiment. But this same struggle is carried on in our cities. There is a frontier life in cities. It is the fiercest of all forms of the struggle for food, shelter, and raiment. More than half the companies that enter this struggle perish, and many a Rose Standish among them. As the struggle is fiercer than ordinary frontier life, it deserves proportionate honor. A light thing to meet the wolves of ordinary frontier life, or any beasts of the forest, compared with meeting the wolves of city frontier life, the dram-shop, the gambling-saloon, the places nearer the pit than either of these, the high rents, the competitions. To 'keep the wolf from the door' is a thing that means more in city frontier life than in ordinary frontier life.

8. It follows from all the preceding propositions that the churches of great cities are now more than ever called to consider the condition of the poor in great cities. If Providence were to write its will on the sky, it would not declare more clearly than it now does, by the leading events of the century, the duty of the church to study the problem of the perishing classes in the great centres of population.

Whoever has not pity enough for the poor to see this duty of the church written in the great and imposing events of Providence in this century, or to regard the occasional discussion, in the churches of our crowded centres of population, of the responsibilities of all other classes toward the poor, as the gospel, either does not know the gospel, or is not a man, but a wolf! (Applause.) What the speaker was about to say would be only the expansion and *application* of the sublime passages of the Scripture just read before the audience.

Mr. Cook then proceeded to consider:

I. *Cheap Homes.* — To illustrate the connections of the moral with the daily physical and social conditions of a population in its homes, Mr. Cook drew a picture of a tenement-house he had visited in Boston, so crowded and dilapidated and unclean that the observance of the common decencies of life was an impossibility in it.

1. The physical health of the poor is their capital in trade; but narrow, crowded, illy-lighted, poorly-warmed, and filthy tenement-houses exert an influence on the physical health so damaging and deadly that the industrial force of a population may be diminished one-half simply by putting the population into such houses.

2. The moral health of the poor is the least withered, and almost the only, source of their happiness: the hearthstone is usually more to a poor man than to a rich man, for it is all a poor man has: moral health in the family is the only fire that can warm the hearthstone; and the most pestilent and deadly of the effects of crowded and neglected tenement-houses is the undermining of the moral health of the occupants.

3. Tenement-houses so crowded and dilapidated and unclean that the observance of the common decencies of life is an impossibility in them, are schools of vice for the children that must be brought up in them.

4. Houses of this class are nestling-places for the city wolves. Intemperance and other of the fierce beasts of the city forest roam abroad through the population in houses of this class, as if through natural and indefeasible hunting ranges; and find there fat prey, (not the fattest, for the richer class oftener furnishes that, but the most accessible and indefensible, because of the disheartening influence of poverty when the moral heat has gone out of the hearthstone).

5. A double price is usually paid by the poor, who are so crowded that they must buy by the parcel and cannot buy by the mass. The poor in crowded tenement-houses in our cities are forced to buy coal by the basket and not by the ton, and wood by the bundle and not by the mass, not only because they have not money to buy more, but because they have no place to store more; and whoever is obliged to buy by the basket or bundle and not by the mass, pays, on the average, a double price.

6. *Unless two conditions can be secured, cheap rents in situations near a city, and cheap railway fares, to and from those situations, the poorest in a city cannot go out of the city to obtain lodgings.*

7. Effort has been successfully made of late in many of the cities of Germany to secure the first of these conditions. The co-operative savings-banks of Germany have been managed on such a plan that large amounts of capital have been put at the disposal of companies of working men for the establishment of cheap homes in the immediate vicinity of great cities.

8. Effort is now being made on a quite large scale in Boston to secure not only the first of these conditions, but the first and second in combination. To combine the two is a peculiarly American idea, which Mr. Josiah Quincy has the high honor of adding to the German plan and bringing before the Massachusetts Legislature.

9. After the repairs are paid for on property rented to the poor, the rent should not be much higher than the legal rate of interest on the amount of capital invested. John Ruskin managed a mass of tenement-houses in London on this plan as an example to England against usurious

rents; and found the occupants, who had paid rents of twenty and thirty per cent, and of whom he required only five per cent, soon able to buy of him a twelve years lease.

10. So far as the solution of the problem of the perishing classes in cities can be promoted by cheap homes, low rents and cheap railway fares, it is evident that the sooner a city gives its attention to the problem the better.

11. This city is growing so rapidly that it may be said that the sun has not risen here on any morning for a year without shining on some new roof. As many buildings have been erected or remodelled here during the last year as there are days in the year. Since 1850 the population has increased one hundred per cent. Since 1865 it has increased thirty-six per cent. The city has yet a village-like character in the connection of plats of land with the great mass of its dwelling-houses. *It is endlessly important that this character be, as far as possible, preserved; and that pestilent tenement-houses be kept out from the first by following the best models in the construction of lodging-houses, but especially by doing everything to promote the establishment of cheap homes.* Mr. Cook commended this subject to the serious attention of the Board of Trade.

II. *Boarding-houses.*— In discussing this part of the topic, Mr. Cook made two preliminary remarks:

1. This city cannot imitate the corporation boarding-houses of Lawrence and Lowell, for there are no corporations here, and the manufacturing population is more largely floating than it is in the cities named.

2. A careful attention to the subject of boarding-houses is seen at Lawrence and Lowell; and that careful attention may be imitated, and ought to be imitated, in this city, until a plan has been found adapted to the peculiar circumstances of Lynn.

Mr. Cook then proceeded to quote ten points from printed regulations in force in the corporation boarding-houses of Lawrence:

1. "No tenement will be leased to persons of immoral or intemperate habits; and any tenant who, after occupancy, shall be found of such habits, or to receive boarders of such habits, will be notified to vacate the premises."

2. "A suitable chamber for the sick must be reserved in each boarding-house, so that they may not be annoyed by others occupying the same room."

3. "Males and females must not board in the same house, without special permission from the agent."

4. "The doors of the boarding-houses will be closed at ten o'clock, P.M., and no one admitted after that time, unless some reasonable excuse can be given."

5. "Nor must company be had at unreasonable hours, and in no case after twelve o'clock at night."

6. "If any boarders are rude or disorderly, or attend improper or disreputable places of amusement, their names must be reported at the counting-room, and they will be discharged."

7. "A proper observance of the Sabbath being equally essential to the formation of good habits and good morals, the preservation of good character, as well as the maintenance of good order, all persons in the employ of this company are expected and required to have some

regular place of public worship in town, and to be as uniform in attendance thereat as possible. In every church in town seats will be furnished gratuitously, if desired, on application on the Sabbath to the sexton of the parish; and such seats may be regularly occupied, till notice is given to the contrary."

8. "On no account must company be received at the boarding-houses during the hours of public worship; nor at any time on the Sabbath will rudeness, disorder, or games of any sort be permitted. Persons holding subordinate positions in these mills, who neglect this regulation, need not expect promotion when vacancies occur."

9. "The above regulations are considered as a part of the contract with each person entering the employment of the —— Cotton Mills, and the Company will require a strict compliance therewith."

10. "I hereby assent to the regulations within, a copy of which I have duly received," is the form of printed words endorsed as a part of the contract, with blanks for name, room, date, and a witness of the signature.

It is exceedingly significant that, after an experience of fifty years, these regulations as to boarding-houses have been found *necessary* in the cotton towns, where one of the most usual forms of the contract with operatives, as appears from the published rules of the mills at Lawrence and Lowell, is, that the contract holds for one year. Contracts with operatives in the shoe towns are so far from being made on this plan, that thirty-three per cent of the operatives here are changeable within the year. It can hardly be supposed, therefore, that if the regulations just quoted are needed in the cotton towns, any *less* careful ones are *needed* in the shoe towns, the floating population is likely to be so much larger in the latter; although it is this very circumstance, and the fact that the shoe factories are individual enterprises and not corporations, which prevents any system of corporation boarding-houses from arising in the latter. It is very important to notice what the *usual* definition of a respectable boarding-house is. Daniel Webster, in a legal case where an agent had lost a large sum of money by not depositing it in a safe, told the jury that the man did not take the *usual* method of keeping the money safely, and over and over repeated with emphasis that the *usual* method had not been taken, until the jury were saturated with a conviction that not to take the *usual* method in such a case was carelessness, justifying the legal action brought against the agent. The *usual* method in respectable boarding-houses, where both gentlemen and ladies are received, is to have two sections of the house, more or less defined, assigned to each class, and not to sandwich rooms of all sorts together, or to lease a vacant room to any one as soon as vacated, especially not to any stranger. When the *usual* method of respectable boarding-houses is followed, Mr. Cook had no objection to boarding-houses where both sexes are received; but it was notorious in this city that the women were often obliged to take the poorest rooms at the boarding-houses. There ought to be two drawing-rooms in each boarding-house of any considerable size, — one for gentlemen and one for ladies, — a bath-room, and a room that could be given at any time to the sick.

Mr. Cook then proceeded to offer six reasons why the Woman's Union for Christian Work, one of the established charities of the city, should have funds placed at its disposal for the erection of a boarding-house.

1. The Woman's Union assert, in their last published Report, that "multitudes" of women, "mostly young, who come from distant homes to engage for a few months in the year in some branch" of the vast trade of this city, are "unable to obtain here comfortable, or even respectable, boarding-places." (Report of 1870, p. 12.)

2. They further assert that, on account of the reason just given, "multitudes" of women, "mostly young," "separated from home influences and home joys, are almost driven, from the nature of things, to seek diversion in the streets, or at undesirable places of public amusement" (p. 12).

3. They inform the public that they have "long aimed to have under their control a well-conducted boarding-house for young women" (p. 13). This enterprise is likely to fail for want of funds. It is a shame — it is a shame which it will require no little benevolence and activity to erase, that, in a city where three or four thousand depend on boarding-houses, the enterprise of this noble charity to open a boarding-house that might be a model for others in the city, should fail for the want of five thousand dollars, or of ten. (Applause.)

4. The Woman's Union is a charity under management that can be trusted. Its committees are the wives of the capitalists and manufacturers of the city.

5. The Woman's Union is confident that, if the enterprise could once be started, it would be a financial success, and support itself.

6. The success of one excellently conducted boarding-house under the control of this established charity, Mr. Cook thought, might be a means of suggesting the right solution of one of the most complicated problems connected with the floating population of shoe towns, namely, how boarding-houses can be what they should be, when a system of corporation boarding-houses is impossible. Try one boarding-house under the control of this charity. If it succeeds financially, rich men here, separately or in company, ought to establish a second and appoint its regulations; and, if that should succeed, a third, until the supply should be found equal to the demand. Mr. Cook said he had volunteered his remarks for the Woman's Union, and was interested in their enterprise, not only for its own sake, but as a means of threading a needle for the future of Lynn, and of cities having similar problems to solve. — *Semi-Weekly Reporter*, March 22.

LECTURE X.

SUNDAY EVENING SERVICES AT MUSIC HALL.

THE attendance at the tenth of the Music Hall services on Sabbath evening was very large, notwithstanding the weather was of the most attractive kind for promenading, and Miss O'Gorman spoke at Thomson's Opera House. In the galleries every seat was taken and numbers remained standing through the whole service, while the body of the house was quite well filled. In announcing that the usual contribution would be taken, Mr. Cook remarked incidentally: "Except High Rock and Long Beach, there are *none* that I fear as to the size of the audience to-night except the Escaped *Nun*," and hoped that she, for the sake of her theme, would have a large audience. After the usual preliminary exercises, the speaker proceeded to the subject announced, namely: "Steps towards the New Birth; or Recent Science on the Signs of the Vices, and on the Best Methods of Escape from Bad Habits." His discourse was based on words found in Acts xvi. 28: "Do thyself no harm," and Galatians v. 16: "Walk in the Spirit, and ye shall not fulfill the lust of the flesh." The discourse was delivered without notes; and we give our usual sketch of what was said.

Mr. Cook having, on previous occasions at Music Hall, spoken of secret prayer, a constant sense of the Divine Omnipresence, occupation with work directed to the accomplishment of high objects of life, a vivid view of God's unrequited love in the cross of Christ, as Remedies for Bad Habits, devoted himself on this evening to the consideration of other remedies rarely brought before popular audiences, but acquiring more and more importance as points of religious truth which recent Science and Christianity singularly conjoin to enforce.

"There is one human being on the earth," said Mr. Cook, in opening his remarks, "who has educated me as the sunbeam educates the violet, whom I have known nine years, and out of whose presence I never went except as a debtor for moral elevation. I had rather touch the little finger of such a soul, than be swept over by whole Atlantics and Pacifics of stained waves. There is another human being now on the earth, a wholly different nature, the presence, or even the memory of whom, is to me as the sound of the trumpet to the soldier in battle; has been this for years; and is likely to continue to be this for all years to come, whether I am on sea or land, at home or abroad. Do not think I am mentioning what is too sacred to be mentioned publicly. These are outlines of friendships and friendships simply. There *is* one human being *not* on the earth — "

1. Mr. Cook then, without finish-

ing the latter sentence, proceeded to a rapid consideration of the power of *high and rich and pure affection in all its forms* as the first remedy for bad habits. Blessed is love; for it is the antagonist of every form of coarse passion. The duty of giving free scope to pure and high friendships, to all the affections of home, and to every worthy and rich affection, and all this as a remedy for bad habits, was here inculcated in strong terms.

2. *The superiority of bliss to pleasure* may now be considered an established doctrine in the fixed portions of the knowledge science possesses of human nature. By bliss, is to be understood the delight of all the rich and high and pure affections; by pleasure, what the world calls such. Three points are pushed into great prominence even by infidelity in the most recent theories of human nature: (1.) All bliss in the faculties arises from their activity; (2.) The moral faculties have a natural supremacy in the mind; (3.) The higher any faculty, the greater the bliss of its exercise. It is one of the commonest, and one of the most significant assertions of physicians, that the pleasures of the vices are a thousandfold overrated by the young, by imagination previous to experience. This is true of intemperance. It is true of high living. It is true throughout the whole range of the vices, mental as well as physical. It is a circumstance of the deepest significance that the infamous portions of the populations of great cities like London, Berlin, and New York, are commonly estimated by the authors who have written the most elaborately on this subject, to be changed once in about seven years, one-half by death under the diseases which are the great red seal of the Lord Almighty's wrath, but one-half by *disgust*.

3. The third remedy suggested was *a distinction between the artificial and the natural passions*. Mr. Cook drew attention to this distinction as a point of great importance in recent science and illustrating scores of passages of the Scriptures. How are we to know what appetite is natural, and what artificial? Is there a test which is itself not artificial, but which any honest man may apply? *A natural appetite*, Mr. Cook held, *does not increase its force on gratification; while every artificial appetite does increase its force on gratification. The two are to be distinguished by this broad difference.* In the appetite for food, gratification of the natural appetite, carried through a lifetime, does not make that appetite stronger. But, in the artificial state of that appetite, induced by the vice of gluttony, by high living, or by dyspeptic disease, the gratification of the appetite only increases its force. The same is true of thirst that is true of hunger. The distinction between the natural and the artificial state of the former appetite is of endless importance in explaining the sorceries of intemperance. The gratification of thirst in its natural state, and this for a lifetime, does not increase its force. The gratification of it in its artificial state, and this for only a year, a month, or, in some cases, for only a day, so increases its force that this appetite alone is found fully competent to ruin both body and soul. This distinction between the natural and the artificial appetites is to be applied through the whole range of human nature.

4. *Abstinence from any physical and seated habit, in order to be successful, must be total; and this not only in act, but in imagination.* This is the organizing principle of the best recent systems of remedial treatment for the intemperate. Dr. Day, formerly of the Binghampton Inebriate Asylum, and now of Massachusetts, makes a rigid adherence to this medical precept the first object to be secured in the case of every patient. Dr. Maudsley's recent investigations as to the effect of alcohol upon the brain show that the microscopic cells of that structure are changed in size, shape, and color by the habit of intemperance; that these changes endure after the habit has been broken off; and that there are physical reasons in the changes why one who has had a seated habit of intemperance cannot reform, except by an abstinence that is total. Dr. Day states that he witnessed the dissection of the brain of a person for many years in practice of total abstinence, but once an inebriate, and found the microscopic cells still in the weak and unnatural state produced by the earlier indulgences. Thousands of the intemperate, who make the most heroic effort to escape their bondage, fail to do so because they do not begin far enough back. It is the doctrine of recent science that there are physical reasons why abstinence, to be successful, must be total; and that it is necessary to resist, not merely the first glass, but the first *thought* of the first glass; that is, to lay the check where the Scriptures long ago laid it, namely, on the *imagination*. This principle applies to every seated physical vice. It is found to be the only spear of Ithuriel of temper celestial enough to touch

"The toad squat at the ear of Eve,"

and make the fiend within start up in his true shape.

5. *No one part of the soul is to be allowed to act without the unforced consent of every other part.* It is a portion of the fundamental law of the United States that no one State is to declare war, or make peace, without the consent of every other State. This is also a fundamental law of the republic of the human faculties; and by it is determined what is, and what is not, *natural* action for any one of these faculties. The base of the brain is not to be allowed to act without the consent of the top of the brain. It cannot be that man is so unskilfully made that the natural action of his faculties is discord. But the very nature of vice is, that it introducs South Carolinas into the republic of the soul; and these declare war or make peace without the consent of all the other States. No vice can carry a unanimous vote of all the faculties. Any action of the soul that does not carry a unanimous vote of the faculties is unnatural. The soul is made on a plan, as much as the eye, the ear, or the hand. *The soul acts naturally only when its action produces continuous joy in all the faculties. No joy is natural that is not full.* The adoption of the principle that no basilar faculty is to be allowed to act without the consent of the coronal faculties, would uproot vice in any soul. Let there be no South Carolinas among the faculties, and there can be no civil war.

6. It is of incalculable worth as a remedy for bad habits, to fasten on the mind, what the advances of recent

science so startlingly demonstrate, that *no habitual vice can be permanently concealed.* In discussing what recent science teaches as to the signs of the vices, endless care is to be taken on two points :

(1.) A suspicious temper is to be thrust aside ; for, at the best, mistakes in reading character will be made. Perhaps a percentage of ten in a hundred judgments will be mistakes at the very best. Signs of the vices, and of diseases or of hereditary taints, must not be confounded. The pasteboard complexion of the student, and the lustreless complexion induced by certain physical vices, must not be confused with each other. The flurry of nerves unstrung by hard work must be distinguished from the flurry of nerves unstrung by dissipation. The depression of spirits induced by dyspepsia or nervous weakness must not be taken for the depression induced by physical vice. But the percentage of ten mistakes in a hundred judgments is a fearfully small one ; and these distinctions are not very difficult to learn to read.

(2.) Only those rules for judging character are to be relied on which have commanded universal assent. No fanciful modern system, however much truth physiognomy or phrenology may contain, is to be leaned on with the whole weight, unless in parts which belong to what all men in all ages have admitted. Nothing less severe than this will serve as just to others, in a matter so delicate as the judgment of character ; nor as safe for oneself, when one depends for preservation from being eaten up alive on this woodcraft of reading character in the forest of human souls.

In many of the vices, nothing is so much feared as exposure ; and, therefore, a discussion of the signs of those vices has an important tendency to repress them, or at least drive them into the dark. The signs of the vices ought to be discussed publicly.

Throwing aside everything fanciful, enough remains to prove that no habitual vice can be permanently concealed.

1. The large signs of the vices every one can read. As to the final signs of intemperance and sensuality, carried to the most advanced stage, all men have in all ages and nations had one judgment.

2. But, in the nature of things, the fine signs must precede the large. In the nature of things, the sun cannot have reached its position at noon without passing through the position of ten o'clock, and of seven, and of four, and of the first auroral flush. The smaller and finer signs of vice *must,* therefore, exist in the countenance. *They are there.*

3. If the signs are there, possibly they can be read. Possibly by patience and skill and practice, an expert, a medical man for instance, can read the signs much further back than ten o'clock.

4. It is universally admitted that the eye has a different expression for every different mood. It is only to make the same statement in another form to say that it has a different expression for every bad mood, as well as for every good. Every emotion — love, fear, anger — brings its peculiar light to the eye ; and that light is the same for the same emotion, in all men and all places. Let any one resolve that for a month he will see no marked expression of the eye without endeavoring to ascertain what it means.

Take the lights about the meaning of which there can be no mistake. Fix these in this study as copies with which to compare other lights. The same light means the same mood, the world around. If once you catch a marked light, and can be sure what it means, when next you catch the same light — if it can be said that the eye ever has twice precisely the same light — you may know what the second light means. If a man feels sure of the meaning of not more than ten or twenty lights, they will teach him much.

5. Complex moods exhibit themselves in complex lights. But when one has learned to read the eyes in ten or twenty lights, taken separately, he can learn to read them in combination.

6. A masked mood or a forced mood has its own peculiar light in the eye. This can be read. One of the most decisive signs of a mask in any person is that the features do not all tell the same story. Want of *consentaneous* expression in look, lips, intonation, gesture, has always been regarded as one of the most decisive signs of a mask. The attempts by the individual wearing the mask to force this consentaneous expression, destroys that sense of *ease of mood* which is one of the first signs of sincerity; and the attempt can be read.

7. Signs of the vices and signs of the virtues will often be mingled not only in the eyes, but in every part of the countenance; and difficult as it is to read simple moods, it is yet more difficult to read complex ones. But when one has learned to read simple moods, so as to recognize them wherever seen, one can learn by patience to read these moods when mingled with each other.

8. The lips are commonly admitted to have a different set for every different mood. But this is to say that they have a peculiar set for bad moods as well as good.

9. After the thirty-fifth or fortieth year, those moods which have habitually been predominant in a man are usually marked on his countenance.

10. Like induces like. The induced mood is a test of very great significance. The mood any person concerning whom you have no strong prejudices for or against, habitually, not for once or twice, but as a plain general rule, induces in you, is an indication of the predominant mood of that person. Woman is perhaps more skilful than man in applying this test, as she is more sensitive to spiritual atmospheres.

11. *Like reads like.* The generous man reads the generous; the just man the just; and so, also, the deceitful the deceitful; and the sensual the sensual. It is, therefore, the man who has most of human nature who best reads human nature. Older reads younger. Past personal experience in the observer reads the same personal experience in the observed. You think you can easily read your younger brothers. But for most of you there are several persons in the world to whom you are the younger. It is the principle that like reads like, which is at once the most searching in its applications and the most universally admitted to be true. It is notorious that a person with a particular illness, or no more than a particular personal blemish, is peculiarly keen-sighted as to the signs of that illness or that blemish in others. *There is vice*

enough in the world to make men keen-sighted as to the signs of the vices.

These principles as to the judgment of character, Mr. Cook said, were those he had been compelled to adopt for himself. They contain nothing fanciful, nothing not universally admitted, and nothing, therefore, on which a man in the collisions of life may not lean. But they are all only a reiteration of the words of Scripture, which are the words no less of human experience in history than of modern science: "Be sure your sin will find you out."

The concluding passage of Mr. Cook's address was as follows:

"There are many Saxon faces in this audience. The blue eyes, the white forehead, the blonde cheek, the fair hair, are signs of the Anglo-Saxon lineage. That race rules the world to-day. It may not always rule it. It rules it for a cause. That race has given to us Goethe and Milton and Shakspeare; and Bacon and Kant and Hamilton and Edwards; and Cromwell and Washington and Lincoln. It wrote Magna Charta, the English Constitution, the Declaration of Independence, the Constitution of the United States. It has bridged the ocean with its commerce, and traversed it with its electric wire. That race, in its German forests, was noted for nothing so much as the spotlessness of its private morals. While yet barbarian, that race, as the Roman historians state, buried the adulterer alive in the mud. The adultress was whipped through the streets. '*Non forma*,' says Tacitus, '*non aetate, non opibus, maritum invenerit.*' 'Neither beauty, nor youth, nor wealth, found her a husband. They considered,' says Tacitus, 'that there was something divine in woman, and that presaged the future, and they did not scorn her counsel and responses.' Youth were taught chivalric notions of honor. Out of this race sprang chivalry. It is this race which has proved itself, in the hurtling contests of a thousand years, both in peace and war, superior to all relaxed Italian and French tribes as the leader of all the world's civilization. The purity of the tribes in the German forests prophesied their future. The hiding of the power of the Anglo-Saxon race has been in the fact that it was at the first free from the sin of Sodom and Gomorrah. That race is passing the trial of power. It is passing the trial of luxury. In the German forests, our fathers, as the Romans found them, were as a race as pure as the dews the forests shook upon their heads. That race has predominated in history because free, even when barbarian, from what elsewhere has been the commonest leprosy of barbarism. It will continue to predominate if it continues free: *otherwise not.*"

At the close of the regular address, Mr. Cook read the following request, which had appeared in the city papers of Saturday, and had received some two hundred signatures from working men, and to which, had time permitted, it was thought it would have been easy to have obtained five hundred:

"*To the Rev. Joseph Cook and the Committee having in charge the Sunday Evening Services in Music Hall:*

"The undersigned, working men of Lynn, having understood that you are about to discontinue the Sunday evening meetings in Music Hall, and being firmly persuaded that the services there held have not only been instructive and of much value to the vast assemblies gathered for public worship, but are likely to become the means of lasting benefit to the whole community, most respectfully, but earnestly, urge that you will, if consistent with other plans and engagements, continue those meetings through the month of April."

Persistent efforts had been made to turn the working men against the Music Hall services; but here was a most urgent request from them, asking for the continuation of those services. Here was a petition from the abused, the slandered working men!

Mr. Cook said he rejoiced to see this request, for two reasons:

1. It was an indication that Sabbath services had been found attractive by a large class in this city, embracing not a few of its most active minds, but not usually found in the churches.

2. It was an indication of value, as showing that the Golden Calf is not Caesar. Miss Phelps, of Andover, had just written a work of fiction on the abuses of the Factory System in the cotton towns. It discussed, with the searching force of that genius which has made the "Gates Ajar" a fireside word in England, as well as America, the state of a factory system, which, as compared with the present condition of the shoe factory system, — the birth of ten years ago, — has the advantage of having had expended on its adjustment the care and experience of two hundred years. It discussed, with a vividness and force which could come only from personal observation, the Social Relations of Employers and Employed, the Employment of Children in Factories against Law, Foul Mouths in Factories, Corporation Boarding-houses, Tenement-houses, Wages, Strikes, Hours of Labor, Relief Societies. Was a committee to be sent to Andover, to prevent the daughter of Professor Phelps and the granddaughter of Moses Stuart, from discussing one of the most vital and complicated of all the problems of New England life? Massachusetts had organized a Bureau of Labor, and paid heavy salaries to that Bureau, to discuss that problem. Elaborate investigations into factory life had been ordered under law. Was a committee to be sent to Boston to stop the action of this Bureau? England had sent forth, during the last hundred years, parliamentary commission after parliamentary commission, to gather, under compulsory legal processes, facts as to abuses practised in the factory systems of its great centres of industry. Was a committee to be sent out to shut the doors of Parliament and of Exeter Hall?

The thirty, forty, and fifty thousand of the future operative population of this city have important interests in the keeping of the capitalists and manufacturers now organizing the new system of labor in this city; but those interests are also largely in the keeping of the working men. Mr. Cook advised the working men to continue efforts to secure such workroom arrangements in the largest trade of the United States as should make the moral perils of themselves and their children as few and small as possible.

As to the continuance of the meetings, the plan of the ten had been a high success. It was now crowned with a peculiar successs in this request for their continuation. April, however, was not February. The usual plan was not to continue such large meetings in our cities beyond March. But, at the request of the working men, there would be an eleventh meeting announced; and this would occur April 2d. — *Semi-Weekly Reporter*, March 29.

LECTURE XI.

SUNDAY EVENING SERVICES AT MUSIC HALL.

Another remarkably large audience was present at Music Hall on last Sabbath evening, although the services, of which this one was the eleventh, have now been continued into a season of the year at which such meetings are usually closed for the season. Every seat in the galleries, and nearly every seat in the body of the house, was taken, and numbers were standing in the galleries and on the lower floor through the whole service. In spite of the suppressed applause which was given as the speaker came upon the platform, and the open applause which occurred at five or six points of the address, it should be put upon record that the audience was of the best quality, and represented well nearly every class in the city.

In announcing the date of the twelfth of the series of Music Hall meetings, Mr. Cook spoke a few minutes on the need of a City Missionary in Lynn. There had been proposed for next Sabbath evening a gathering of all the churches in the city to consider this subject. Mr. G. P. Wilson, City Missionary of Lawrence, was to address this assembly. Mr. Wilson was one of the most successful city missionaries in eastern Massachusetts. At the time of the falling of the Pemberton Mills, his services to those who suffered by that disaster in Lawrence were invaluable. A city missionary would cost Lynn, Mr. Cook thought, some two thousand dollars a year. But he would probably save the city at least half that sum every year. It was found, in the experience of city religious charities, that by timely aid to the poor, and by just discrimination between real and pretended causes of want, a most important addition can be made to the number exempted from all necessity of depending on the overseers of the poor or any other charity. In Lawrence, so true had this been found, that Mr. Wilson had been appointed to a position in the city government as disburser of the charities of Lawrence; and this as a measure of economy. It was a shame to bring forward the idea of the expense of two thousand dollars a year, as an objection to the appointment of a city missionary in a population of thirty thousand people, with twenty churches to unite in contributing this salary. But if such an objection could have weight with any, Mr. Cook called attention to the act of the city government of Lawrence in using the city missionary as an instrument of economy in public expense. It was time and high time for Lynn to have a city missionary. Of the eight or ten cities of nearly the size of Lynn this side the Connecticut river, all but two or three had had city missionaries for years. The Young Men's Christian Association and the Woman's Union, so far from objecting to the appointment of a city missionary as an encroachment on their fields, were eager to secure one. The success of the recently established Boys' Mission was one among a multitude of proofs of the need of a city mission. Independently of consultation with each other, the several pastors and churches had been found to be interested in this theme. Two important though somewhat informal and insufficiently attended meetings of the representatives of the churches, had recently pushed the subject forward to the appointment of a committee from all denominations to draft the plan of a Lynn city mission. Mr. Cook purposed to prove his interest in the enterprise of a city missionary by omitting services for one Sabbath evening at Music Hall. The twelfth of the Music Hall services would, therefore, be held April 16th. Mr. Cook most urgently invited the whole audience to be present at the city missionary meeting at the Common Street Methodist Church, next Sabbath evening.

After devotional exercises, Mr. Cook took up the subject that had been previously announced: "Swindling Public Amusements, the Street School, and Club-Rooms in Lynn." The texts used embraced the passages in Rev. xiv. 1–12; and Proverbs ix. 10–18; especially the words: "They have no rest, day nor night, who worship the beast and his image, and whosoever receiveth the mark of his name."

"Give me a union of the Pulpit, the Parlor, the Press, and the Police," Mr. Cook began his address by saying, "and I have all I want in respect to any moral reform. But, give me a union of the Pulpit, the Parlor, and the Press, for any cause in regard to which legal enactments are expedient, and I will gain support of the Police. Give me a union of no more than the Pulpit and Parlor, and I will in time gain the Press; and, with the three, the Police. Give me only an unanimous union of the Pulpit, time enough, and a cause of importance, and I will gain the Parlor; and through the two, the Press; and through the three, the Police." It is greatly important to notice this order in which moral reform usually progresses. Pulpit and Platform, which are so far analagous in their means of influence as to be ranked together, constitute one of the four greatest powers of modern civilization. The Parlor is a second; the Press a third; Public Law a fourth. Without affirming that any one of these powers is greater than any other, it may yet be affirmed that, in the nature of things, the order of progress for a moral reform, like anti-slavery or temperance, for example, is from the first to the fourth, rather than from the fourth to the first. History shows that, in free communities, in the majority of instances for the last two hundred years, moral reform has passed from the Pulpit and Platform to the Parlor; and from these two has conquered the Press; and from these has won the recognition of Public Law.

From these circumstances, Mr. Cook went forward to draw an inference as to the importance of preserving the Pulpit and Platform from bondage. It is this natural and irreversible order of the four great means by which the interests of every class in society are protected, which makes it of such transcendent importance to every class, and, indeed, to the existence of a free civilization itself, to beware of all influences prejudicial to calm, guarded, dispassionate public discussion, in the Pulpit and Platform, of any topics touching any moral issues of society, even if tremendous interests and tremendous passions are involved.

Mr. Cook went on to illustrate this teaching of history, by speaking of the snare Slavery wove for the pulpit. Social and industrial tendrils, interwoven from the Great Lakes to the Gulf, were the meshes of the web that choked a great part of the pulpit of the North. Slavery must not be discussed, it had been said only fifteen years ago — although the day seemed a century back, by the way we had forgotten its lessons as to this bondage of public discussion! — lest these industrial and social tendrils, interwoven from the Lakes to the Gulf, should be made to rasp on each other and to bleed.

The day had not yet come in New England; but it had come in Old England when the relations of Capital and Labor, although calling out the most vigorous parliamentary legislation, weave a snare for public discussion; and for the Pulpit more than for the Platform. The meshes of the snare are not as heavy as those Slavery wove; but we are to beware, in the manufacturing centres of New England, where the distinctions between rich and poor are wide, of the weaving of even the first lines of that web.

Massachusetts has found industrial and moral questions connected with the great centres of her manufacturing populations so important as to appoint a Bureau of Labor, with heavy salaries, to investigate, under all the power of law, these problems, sure to become as complicated in New England as they have become in Old England. Already there exist in New England points connected with factory life — the employment of children in factories against law, for example — where the meshes of the snare have been woven so strongly that the officers of this Bureau of Labor publicly report to the Massachusetts Legislature that the power of State law is hardly sufficient to break the lines of the web. It is of great public importance to preserve the Pulpit of the manufacturing centres of New England from a bondage into which it is not to be charged as yet with falling; but into which the pulpit of England, from a fear of offending wealth by discussing the interests of the poor, or a fear of offending the poor by discussing the interests of wealth, has largely fallen.

It was in view of the extreme importance of these circumstances that Mr. Cook said that, in introducing, as he was about to do this evening, new points involving something of criticism, he wished to begin by thanking the audience, and through it the city, for the thorough manner in which it had vindicated from bondage the interests of public discussion in the pulpit of one of the most complicated of the problems arising in one of the most important of the manufacturing centres of New England, and connected with the interests of the largest trade of the United States. In view of the public importance of these considerations, Mr. Cook said: "I do not depend upon my salary for a living; but, if I had so depended; if I had wished, as I never have, for a situation in this city; if I had had a young family depending on me; and if I had had a timid wife holding me back, I should have done precisely what I have done." (Continued applause).

Mr. Cook then proceeded to discuss Swindling Public Amusements, with especial reference to the Theatre.

1. It is a fact of current history that in a majority of the theatres in the most moral cities, scenes are admitted which have on them so much of the mark of the beast that they could not be photographed without the pictures being actionable at law. Mr. Cook drew a distinction between the higher theatres and the lower; but called attention to the circumstance that the highest admitted at times scenes bearing the mark of the beast, and had done this repeatedly in Boston within four years. In proof of this assertion, Mr. Cook alluded to an article published in May, 1869, in the *Atlantic Monthly* — a most excellent authority in art, and a most liberal one in theology! — written by the accomplished editor of the *Monthly*, Mr. Howells, and running a red, crooked thunderbolt through the plays having on them the mark of the beast. "This English burlesque, this child of Mr. Offenbach's genius, and the now somewhat faded spectacular muse, flourished the past winter in three of our seven theatres for months, five, *from the highest* to the lowest, being in turn open to it." This is Boston, where the theatre is perhaps less open to criticism than in any other city in the world.

2. It is a fact of past history that the theatre, in spite of all attempts to defend it, has always had, on the whole and on the average, a bad reputation; and had this, on the whole and on the average, even in Pagan antiquity. This permanent reputation of the theatre century after century is a fact of extraordinary significance. As one out of a score of similar illustrations, take the circumstance that in Rome, in 165 B.C., Publius Scipio Nasica, ordered a theatre, which had been partially erected, to be pulled down, exclusively on the score of public morality.

3. In view of the manner in which exceptions to this current and past reputation of the theatre are spoken of, the exceptions are themselves a proof of the rule. It is most important to recognize the exceptions. Mr. Booth is pointed out as an exception. But when it is *common* to hear exceptions spoken of *as such* in regard to any profession, *the admission is involved that the general rule is the opposite of the exceptions*. Suppose that in the medical or clerical profession it were *common* to hear men of a certain character spoken of as exceptions, what would the inference be as to the general rule?

4. *The theatre is outgrown as a mirror of the current social life and political ideas of a people, having, in both these capacities, which once gave the theatre importance, been superseded by the lyceum, the growth of modern literature, and especially by the immense increase of the power of the newspaper press.* Mr. Cook urged this point at length. In Shakspeare's day the theatre was what it could never be again, as a mirror of the current life of the people. If a stranger were to come to Boston to-day to study New England, he would find the lecture-room platforms a better mirror than the theatres of the most important phases of the current social and political New England life. It was notorious that as an instrument of political influence the theatre had

been superseded by the lyceum, the newspaper press, and the growth of modern literature; and though it was not equally notorious that as a mirror of current life it had been superseded by these agencies, Mr. Cook thought that the latter proposition was now, and was likely to remain, equally true.

5. *It is safe to assert that the problem of reforming the theatre, is rendered more difficult than ever before, now that the office remaining to it is almost exclusively that of amusement, its two former offices, as mirror of the life of the people and as a political instrumentality, having been superseded.*

6. It is very important to notice that, even as a source of amusement, the theatre is now much more largely superseded *for the cultivated class*, than it was in Shakspeare's day. The theatre, even as a place of amusement, cannot be again what it was, since other forms of amusement, for both cultivated and uncultivated, have increased perhaps fifty-fold in the last two hundred years. The newspaper press, the lyceum, the growth of illustrated literature, libraries, the popular use of music, are all increasing in power of amusement; and their rivalry adds difficulty to any attempt by the theatre to succeed as a place of amusement; and it must succeed as that, or not at all.

7. *Amusement is now so open elsewhere to the cultivated that it is found that a theatre managed according to their tastes as a place of amusement does not pay the expense necessary to satisfy a cultivated taste.*

8. The temptation to address, in the theatre, the tastes of the uncultivated and the vicious, is, therefore, extremely great; but it is a temptation arising from circumstances of permanent and increasing influence in modern civilization.

9. In European society, especially in the French city populations, the condition of the mass of the people, is politically, socially, and intellectually, so different from the condition of American society, that the former can be amused by what is not at all adapted to the latter. *We are superior to the French plays.* But it is from abroad that the plays which do most to corrupt our cities come; and, instead of being admired as foreign, they ought to be criticised as such. *Theatrical amusements fitted to satisfy the mass of a European population, are, because of that fitness, inferior to the demands of an American population.*

10. The low theatres are a host, and the better only a handful, and the latter have much less influence than the former. But of the latter it must be said that to-day, in Boston, where the purest theatre of the world exists, enough of the mark of the beast is admitted into the plays to make it difficult for a man to go in at the doors of the theatre without stooping; and into the low theatres a man cannot go except on all-fours.

Dr. Holmes remarks that every man leads or is led by something that goes on all-fours. It may be added, that if the human race is ever to find a paradise, it will not enter it on all-fours.

Mr. Cook then proceeded to discuss the Street School, with especial reference to the corrupt illustrated popular press. This he termed the Leprous Press — the Sewerage Press.

1. If it must be acknowledged that its force is increasing, it must be ad-

mitted also that the force of the better press is vastly increasing.

2. European, especially German and English travellers, point to the failure of our city governments to clear the street windows of the leprous and sewerage press, as a proof of the inferiority of democracy as a form of government. Boston and New York allow what would not be allowed in London.

3. The offices from which most of these sheets emanate are usually of the shabbiest and lowest kind. Mr. Cook described the office of one of the leprous sheets of Boston he had visited — four broken panes in the front door, a hall full of filth and broken furniture, and the editorial room only a small space cleared in an apartment filled elsewhere as a storeroom of furniture, and containing nothing belonging to the editors that could not have been packed up in an hour. In New York he had seen a basement-room used for similar purposes where a tar-can was an ornament on the walls, and the only counter two boards laid on two barrels. The offices are like the Arab's tent, capable of being packed up at the shortest notice; and this is one reason why the police have difficulty with them.

4. It is worth much to drive these papers into such a position that they can be read only on the sly; and this, a public statement of their character helps to do.

5. Mr. Cook took the responsibility of advising the public not to patronize, for anything whatever, newspaper dealers in the leprous press.

The speaker referred to the habit of evening walking on the street, somewhat characteristic of this city, with its beautiful scenery close at hand, and pointed out that the perils connected with that habit have many illustrations elsewhere.

It was Mr. Cook's intention to have spoken on the Club-rooms; but he had addressed the audience, wholly without notes, for the full hour, and would take up that topic at the next service. He had been with two policemen through two of the club-rooms of the city. He had, of course, exceptions to make as to social and literary clubs. He knew of one club that did *not* open Sundays. But, as to the mass of the club-rooms, they were places where Satan gave his degree of LL. D. What does that LL.D. mean? Liquor and Locked Doors. (Applause). — *Semi-Weekly Reporter*, April 5.

LECTURE XII.

SUNDAY EVENING SERVICES AT MUSIC HALL.

THE twelfth of the series of Music Hall services was attended by another great audience. The galleries were crowded, hardly a seat was vacant on the lower floor, and perhaps a hundred remained standing through the whole service, which was two hours and a half in length.

Mr. Cook announced that he wished to introduce, for two of the three hymns of the evening, congregational singing; and to consecrate the spaces of the hall by the voices of the whole audience rising and uniting in devotions. It was only recently that the grand choral congregational singing of the German churches had begun to be introduced into the American. The whole audience rose in the galleries and on the lower floor, and sang together with the most impressive effect, "All hail the power of Jesus' name." At another point of the evening, the whole audience rose again and sang together, "A charge to keep I have." The audience was led by the usual select chorus of some fifty singers, Mr. C. E. Ashcroft acting as pianist; and the devotional effect of the union of voices of the audience was of a very marked kind.

In view of the size and spirit of the meeting on the previous Sabbath evening, inaugurating, in Lynn, the enterprise of a city missionary, Mr. Cook congratulated the audience, and through it the city, on the prospect that within a few weeks a Board of Advice would be chosen; and he wished to make a few remarks on the *financial* future of the mission.

1. In several of the city missions in Eastern Massachusetts, the salary of the city missionary is paid exclusively by the great manufacturing corporations. In Lawrence, Mr. Wilson never goes before the churches to ask funds for his own support. The churches are appealed to, and give generously; but these contributions are for the general charities of the mission, aside from the salary of the missionary. This last is paid by the great corporations of the cotton mills. Mr. Cook read from a recent report of the Manchester City Mission statements showing that the great corporations in that city pay the salary of the missionary.

2. In this city there are no great manufacturing corporations. The great factories are the enterprises of individuals, or of business firms; and rich men will be appealed to here to give as individuals, rather than as members of corporations. If the churches expect that the business firms here will maintain the city missionary as the business corporations do elsewhere, Mr. Cook thought that, from the nature of the organization of business here, the churches would be disappointed. A most powerful appeal, however, may be made to business men here, in view

of what business men do elsewhere, to aid a city mission as largely and regularly as individuals as business corporations do in a city where corporations exist.

3. In view of the two preceding propositions, it is evident that in this city a peculiar financial responsibility in regard to the city missionary will rest on the *churches.* It will require a long head to point out where the funds, not only for the salary of the missionary, but for the general charities of the mission, are to come from, as a source on which regular and permanent reliance can be placed, if they are not to come from the churches, and rich men as members of the churches.

4. Mr. Cook thought that a contribution from the churches, equivalent to a dollar a year from each church member, was as small as would be adequate to the purposes of the mission. There were now twenty-six hundred church members in the fifteen evangelical churches. He would estimate the members of the unevangelical churches at one thousand. Thirty-six hundred dollars were not too large an allowance for the salary of a competent missionary and all the multiplied charities of the mission. It was surely not too much to ask of the churches to give annual contributions equivalent to one dollar for each church-member. But, at this small requirement, the First Congregational Church would give two hundred and ten dollars a year; the Friends, two hundred and forty; the Common Street Methodist Church, three hundred and fifty-five; St. Paul's, one hundred and thirty-five; the First Baptist, three hundred and forty-four; the South Street Methodist, two hundred and twelve; St. Stephen's, one hundred and thirty-three; the Central Congregational, one hundred and forty-nine; the Maple Street Methodist, one hundred and forty-one; the High Street Baptist, two hundred and ten; the Boston Street Methodist, two hundred and forty.

5. These contributions by church-members, or equivalent funds from other sources, are to be secured by the churches not for one year only, but year after year, and increased in amount as the city and churches increase in size. The difficulty would turn out to be in making a contribution so small as to be equivalent to only a dollar from every church-member of the city, an annual and permanent affair.

6. Such a man for city missionary as will unite the churches is, therefore, a *financial* necessity in respect to the mission.

7. If different denominations are to be united in the enterprise, such a plan of union as will best secure the financial support of the mass of the churches that must sustain the mission is plainly a financial necessity. Mr. Cook said he had taken the position that, after the right man had been secured as city missionary, he should be allowed, as Mr. Wilson had been at Lawrence, and as was the usual plan elsewhere, to follow out his own ideas in the distinctively religious part of his work, under the general direction of the majority or two-thirds of the Board of Advice. This was not a denominational, but a *financial* position. If the relative numbers of the churches in the two sets of denominations which it was proposed to unite were reversed, the speaker said he should have taken

precisely the same position; and this for the same reason, namely, a desire to secure the good of the city by securing the *financial* success of the mission.

8. Mr. Cook concluded his remarks on the City Mission by an appeal to the city to make the financial support of the mission worthy of New Lynn. The City Hall had not been found unworthy of the new growth of the city. The new school-houses and the new churches were not built penuriously. Not a business-block could be erected or a house remodelled in the city, now, without the influence of New Lynn appearing in the scale of expense adopted. The speaker said he had resolved some ten months ago not to leave the city without doing his best to see a city mission organized. Two years ago the local pastors had carefully discussed the theme. The Young Men's Christian Association had more than a year ago gone so far as to correspond with one or two men in order to find an agent for such a mission. Mr. Cook said that, in this agreement of many different parties, wholly without consultation with each other, he rejoiced to find an indication not merely of the great importance of the enterprise, but of the grounds of hope that the financial difficulties it is sure to meet, will be met in a manner worthy of the city. Build the City Mission, as you build everything else, for New Lynn. (Applause.)

Before announcing the usual contribution of the evening, Mr. Cook took occasion to read the following resolutions, in denial of the rumor which had been circulated in certain quarters of the city, for the last few weeks, that the series of meetings at Music Hall had not been authorized by the Committee of the First Congregational Church. The rumor, it will be seen, is wholly false, as the first of the following resolutions had the signatures of every member of the committee, and the second passed by an unhesitating majority vote. The church has sent its janitor and organist to Music Hall; its officers have taken the contributions there; the public prints have contained cards of thanks to the Lynn Choral Union and other singers for services rendered to the First Church and Society in the meetings at Music Hall; and the Treasurer of the Church and Society has eleven receipts in his possession, drawn in his name, for payment of the expenses of the hall. In the appointment of the committee, unusual care had been taken to give it "full power" to provide for "all" the meetings of the week, including a *second service*, and especially a sabbath evening prayer-meeting and preaching-service, both of which were included in the first meeting held at Music Hall, although the prayer-meetings had been discontinued after the first meeting, on account of the inconvenience arising from the size of the audience. There had been some differences of opinion in the committee; but they had all signed the paper providing for the ten meetings; and it would be noticed that the reasons given for the continuance of the meetings, in the second resolution, include all those given in the first, with others of weight, and are all matters of fact and not of opinion. The committee, having finished its duties, had been discharged; but nothing in its action had been criticised in the discharge; and provision had been made by the committee for meetings through April. The thirteenth and last of the second services of the First Congregational Church, at Music Hall, Mr. Cook announced for the last Sabbath in April, this being the last of the Sabbaths for which he had made engagement to labor in this city; and his last Sabbath before beginning his preparation to go to Europe.

" Whereas, the attendance at the meetings thus far held at Music Hall has been extraordinarily large;

" And whereas, the opportunity of usefulness is plainly great;

" And whereas, the expenses of opening the hall have been reduced by contributions from the audience, by subscriptions, and by the pledges of certain parties, so that for ten nights, including the five meetings already held, the expenses to the Society shall be only fifteen dollars a night; and, by pledges of other parties, for four meetings beyond the first six, the cost nothing whatever: Therefore,

"*Resolved*, By the Committee of the First Congregational Church of Lynn, appointed to select for the Church and Society a temporary place of worship, that the First Congregational Church and Society worship at Music Hall for several nights more, announcing, however, but one meeting in advance.

"*February* 13."

" In view of the following circumstances, namely:

" 1. That a petition urgently asking for the continuation of the services at Music Hall has appeared in the city papers, and received the signatures of two hundred working men;

" 2. That for ten Sabbath evening services, running through two months and a half, the audiences have been extraordinarily large;

" 3. That the opportunity of usefulness opened to the First Church and Society is plainly great;

" 4. That no risk whatever is now incurred by the church as to expense, funds having been pledged which cover any risk to the end of April;

" 5. That the alternative would be to hold no second service, or a service much less well attended; no room for prayer-meetings on the Common of the city being open to the First Church since the burning of the house of worship of the First Church; and the season of the year not being that most favorable to the success of devotional meetings;

" 6. That the reputation of the acting pastor has been publicly and persistently assailed from outside the parish, on the most serious points, in a manner which has caused the unanimous passage, by the church and society, of a resolution vindicating their pastor; and in a manner which the success of the services at Music Hall is calculated to repel;

" 7. That it is highly important to the First Church to do nothing to repel young men and the population not usually attending church, both of which classes have been largely attracted to the services at Music Hall;

" 8. That the churches of the city have heretofore united in holding Sabbath evening services in the Lyceum Hall, in this city, when the audiences were more than half less than the audiences now held at Music Hall:

"*Resolved*, By the Committee appointed by the First Church and Society to provide a temporary place of worship for the First Church and Society, that the resolution recently passed unanimously by this committee to hold several more meetings at Music Hall, announcing but one meeting at a time, be extended in its application so as to cover any meetings the committee choose to announce for April.

"*March* 30."

Mr. Cook then took up and spoke for an hour and a quarter on the subject which had been announced for the evening: "Club-Rooms in Lynn, Gambling-Rooms, and Dwelling-house Sabbath Drinking Parties; or the Duty of making Christian Public Sentiment in Cities a Police behind the Police." On account of the importance of this subject to the city, we wish to report the discussion of it more in detail than we have space for in the present issue. A full analysis of the address will therefore appear in next Saturday's *Reporter*. — *Semi-Weekly Reporter*, April 19.

We give, to-day, our promised report of the address delivered by Rev. Joseph Cook, last Sunday evening, before the crowded audience at the twelfth of the Music Hall Sabbath evening services. The address, which was delivered entirely without notes, was listened to with deep attention throughout, applause being bestowed at several points. The text was from Ecclesiastes ii. 13: "I saw that wisdom excelleth folly, as far as light excelleth darkness."

After an introductory allusion to John B. Gough, who had spoken in the Hall the previous Tuesday evening, Mr Cook said he should divide the topic which had been announced into four principal parts: Open Bars, Club-Rooms, Gambling-Rooms, and Dwelling-house Drinking Parties.

I. *Open Bars.*

1. "Standing in one of the open squares of this city last evening with a policeman," said Mr. Cook, "I said: 'There is a white horse in the centre of the square. If you stood where that horse stands, into how many places where liquor is illegally sold could you throw a stone?' The policeman, a cautious man, replied, looking about the square: 'One, two, three — seven — ten — twelve or fifteen, at least.'" The day might come yet, in the future of this city, when, unless public sentiment in a larger population is made a police behind the police, ten or fifteen such squares may be open within hearing of the waves of Massachusetts Bay! But the evil one such square and all it represents can do is not easily calculated.

2. It is common for the best portions of the Pulpit and Parlor and Press, in our cities, to complain of the inefficiency of the Police in executing Temperance Laws.

3. It is also common, in our cities, for the Police to complain that there is no public sentiment to justify them in the execution of Temperance Laws; that is, they complain of the Pulpit, Parlor, and Press for not awakening such sentiment. Go to the Pulpits and Parlors, and you find complaints made of the City Marshals and Chiefs of Police: go to the City Marshals and Chiefs of Police, and you find complaint made of the Pulpits and Parlors.

4. Both these sets of complaints are just. As a general rule, it is true that the Police are right in

their complaints of the Pulpit, Parlor, and Press; and the Pulpit, Parlor, and Press right in their complaints of the Police.

5. It is impossible for the Police alone to secure the right condition of a great city in respect to Temperance or any other moral reform; and it is also impossible for the Pulpit, Parlor, and Press, taken without the aid of the Police, to secure that right condition.

6. The two sets of ideas represented in these complaints must, therefore, recognize each the just claims of the other, and clasp hands together under the neck of the moral difficulties always arising in crowded centres of population, or the great problem of the right management of cities will never be solved.

7. The scheme of State Police, or a constabulary representing the whole State, and executing the temperance laws of the State, unites, perhaps better than any other plan, these two sets of ideas; for it brings the vigorous temperance sentiment of the rural portions of a State to the support of the less vigorous temperance sentiment of the cities. Unless the portions of New York above the Highlands of the Hudson helped to rule the part below the Highlands, the latter would hardly be governable, except by martial law. The plan of a State Police is one of the best solutions yet invented for the problem of the perishing and dangerous classes in cities.

8. There is nothing undemocratic in the plan of a State Police. The republican plan of government is, that the executive should be coordinate in the sphere of its power with the legislative and the judiciary. The legislative power makes laws, including temperance laws, for all the State. The executive power should execute the laws, including temperance laws, for all the State. This is the true theory of republican institutions, and was the practice in Massachusetts until about fifty years ago.

9. It follows from the preceding propositions, and has now been proved by experience, that a State Prohibitory Law, enforced by a State Police, brings, better than any other plan, the force of the better parts of a State to the aid of the worse. Mr. Cook declared himself in favor of Prohibition.

10. In the most unfortunately situated cities, however, taken without the aid of any public sentiment outside of themselves, there are tides of temperance sentiment enough wasted in the churches and parlors to make a police behind the police, if the Pulpits and Press would only give that sentiment adequate expression.

11. Massachusetts is now, in respect to temperance legal enactments, under what may be called the system of Local Option. Each town has its option whether it will permit the sale of the malt liquors, or not. This city has, at present, a very good set of temperance enactments. But several cities in Eastern Massachusetts have recently been turned against temperance by small majorities in municipal elections; and this with the most ruinous results.

12. *While the system of Local Option prevails in temperance legislation, it is plain that great responsibility rests on the local Pulpit, Parlor, and Press.* On the grounds of these propositions, Mr. Cook made an appeal to the Pulpit, Parlor, and Press of the city

to unite in making public sentiment a police behind the police to aid the police to shut the open bars.

II. *Club-Rooms.*

1. It is important to notice, in cities, that middle ground which exists between respectability, on the one hand, and action which the police can touch, on the other. Men are divided not only into the standing and fallen. There is a third class — the *falling*. This is found in this middle ground oftener than elsewhere. It is peculiarly the province of the Pulpit and Parlor to study this middle ground.

2. Club-rooms in cities occupy, as a mass, this middle ground; although the better of them rise into the region of respectability, and the lowest sink into the region of illegality, and fall into the hands of police law.

3. Clubs increase in power with the increase of the power of cities. This is true of the political and literary and social clubs, on the one hand, as well as of the drinking and gambling clubs, on the other. Wendell Phillips complains that Massachusetts is ruled politically by Boston Clubs; and he was fortunate in being able in a recent political campaign to put one of those clubs into the pillory.

4. When severe temperance legislation is executed in cities, the lower class of club-rooms blossom. Many of the club-rooms of this city had their origin at the time the Prohibitory Law was thoroughly executed here. This is no argument against a Prohibitory Law; but it is an indication that drinking is an essential object with the lower class of club-rooms.

5. The Police in this city often find places that are in reality dram-shops, defended, when brought before the city courts, on the plea by the lawyers that they are club-rooms. This is another indication of the true character of the lower class of these places.

6. Mr. Cook said he had studied the club-rooms of Lynn by the aid of the police. He found, on inquiry of the proper officers, that in Lynn, as in larger cities, it was not unusual or regarded in any sense improper for persons wishing to study the moral condition of the population to avail themselves of the aid of the police; and he paid a high compliment to the police for the courtesy and fulness of the aid they give in such a study of the city. He had been, with company from the police, through four establishments, one of which was one of the worst gambling-rooms in the city, another a club-room of the better class, and two which were called specimens of the club-rooms of the worst class. He had gone in his usual dress without disguise; and in two of the club-rooms had had quite full and always courteous conversations with persons present in them; and he should speak from personal observation, as well as from the information of the police.

7. In the places he had visited, and which were called specimens of the worst class of club-rooms, he found the outer doors locked and provided with a small window through which any who applied for admission, and especially the policemen, were observed from within, before being admitted. It was useless to set up the claim that this arrangement was intended solely to keep out persons not members of the club.

The police found their investigations in these rooms barred by the windows and the locked doors. When a policeman applies for admission, a delay usually occurs, long enough for any evidence of illegal liquor-selling or gambling that may be going on within, to be put out of sight. "If you are all right, why do you not admit us without so much delay?" the police significantly asked twice in one of the rooms. "We are all right," was the answer. In the cases he had observed, Mr. Cook found nothing attractive, but everything shabby, about the rooms. In each case, a bar-room and an adjoining sitting-room made up the establishment. The places are fitted up with no taste, and one of them was so shabby and repulsive that Mr. Cook said he came out of it thankful that he had a clean pillow on which to lay his head. The floor was filthy; the walls and windows dirty; the table bare; and all the furniture in the room would not have sold at auction for fifteen dollars. There was one little colored print, of no significance, on the walls; a cask with a facet (probably containing water!) on the floor near a closet and a bar. In another room he saw three of the most leprous illustrated newspapers on the bare table; and the pictures on the walls were in part of the same class with those in the papers, and in part of boxing and racing scenes. It was the general rule in the city, as he had been informed by the police, that the club-rooms of the lower class are shabby and repulsive places. But they are frequented from morning to night, and till late at night; and most of them keep open on Sundays. The police estimate that there are at present some thirty or thirty-five club-rooms of this class in the city.

8. Where is the money in club-rooms made? Mr. Cook said he regarded this a test question in determining their true character. He had put this question everywhere; and had concurrent information, both from the police and those whom he had conversed with in the club-rooms, that the money is made from the bar and from toll on the table. When a person sits down at a club-room table at a game, it is customary to charge a small sum — perhaps twenty cents — for a sitting; and the money made by the managers of club-rooms is from the bar and from this toll on the games. It used to be said of John Quincy Adams, when in Congress, that he had a carnivorous instinct for the jugular vein and carotid artery of an argument. Mr. Cook regarded the fact that the club-room managers make their money from their bars and the toll on the tables as the jugular vein and carotid artery of the proof that, whatever may be said of the character of these places as opened for social intercourse or amusement, their essential business is gambling and drinking.

9. When the current sets all one way, in any social gathering, the effect of putting oneself into that current is always of a very marked kind. In a prayer-meettng, not of the hypocritical sort, the current sets all one way; and this toward virtue. This is the secret of the power of such a meeting, and is one reason why some men are so shy of putting themselves into that current. But, in a club-room of the lower class, the central currents are gambling and drinking; and these draw in gradually the side swirls of mere social

intercourse or good fellowship, until it may be said that the current all sets one way, and that toward the direction the central current takes. This is the secret of the influence of putting oneself into that current. The thoroughly equipped low club-room of a great city is a prayer meeting of the Devil's Church.

10. Those who, at any age, are under the power of bad habits, and the young, are the two classes most injured among the membership of club-rooms; and both these classes are most numerously represented in that membership. It is not an unknown circumstance for very old persons to belong to club-rooms. The young have so much representation in them that Mr. Cook found that teachers exhibit high interest in the subject. The High School of this city is a gem, and this even on the Massachusetts coast, more irradiated by the gems of the schools than any other coast in the world. The princely Principal of that school, who knows the city well, is not without the information as to club-rooms needed to bar out their influences from the first.

11. The class of clubs meeting for political, social, or literary purposes, Mr. Cook said he did not intend to discuss. He had visited one of the clubs of this class, conversed with its members, and read its constitution, which contained provisions, which the police believed were thoroughly executed, against all liquor on the premises, and against opening on Sundays. He saw nothing in any part of club-life, however, which could afford any adequate reason why a busy man, with but a few hours to devote to his family each day, should give four, three or two evenings, or even one evening a week, regularly, to a club. (Applause).

12. In the low club-rooms, the Sabbath is often the worst day of the week.

III. *Gambling-Rooms.*

1. Arrangements for blockading or retarding the coming in of the police were visible enough at the entrance of the gambling-room, as well as at the entrance of the club-rooms.

2. Mr. Cook described the gambling-room he had visited, and a room connected with it, as of the shabbiest and forlornest character; and the furniture in the gambling-room, aside from a rough table, as not worth ten dollars.

3. And yet, in this room, gambling of the most thorough sort, the police had reason enough to believe, was carried on. It did not require a long apprenticeship to acquire skill enough to shake four small flattened sea-shells, called props, like dice in the hand; and determine a game by the shells coming, four or two, the same side up, and making a *neck;* or three one side up and one the other side up, and making an *out.*

IV. *Dwelling-House Drinking Parties.*

1. When the Catholic clergy of Boston were called before the Massachusetts Legislature, to testify in the License Hearings conducted by Governor Andrew and President Miner in 1867, they testified in the strongest terms that one of the most deadly evils they met in the population they were called most to study, was the sale of liquors in private dwelling-houses. They represented that there were whole streets and squares of dwelling-houses, especially

among the population of foreign descent, where the whole moral ground was as springy with these sales as a marsh is with rills. They condemned this habit of the population as an evil of the first magnitude; and as not altogether confined to the foreign population.

2. Mr. Cook said that he knew a street in this city where he had reason, from information received from the police, to believe that liquor is sold in every third dwelling-house.

3. Pass through such a street on the Sabbath, and although no outward disorder may be seen, it is common to find, sometimes on one street, five or six dwelling-house Sabbath drinking parties. Ten or fifteen persons in some back room of a dwelling-house are gathered about a supply of liquor, drinking. When the police appears, the group scatters, some up stairs, some to the closets, some to the cellars.

4. The police assert that they had rather have four open bars than one dwelling-house drinking hole. — *Semi-Weekly Reporter*, April 22.

LECTURE XIII.

SUNDAY EVENING SERVICES AT MUSIC HALL.

At the thirteenth of the Music Hall services, there was present another great audience. The galleries were crowded; nearly every seat was taken on the lower floor; and a considerable number remained standing through the whole service, which was five minutes less than three hours in length. It should be put upon record that the audience was as remarkable for quality as for size. We noticed persons present from Boston and Salem, who said they had come to the city to attend this service. Mr. Cook spoke two hours and five minutes. The platform, which contained a beautiful collection of flowers arranged on each side of the open space allotted to the speaker, was occupied by a select chorus of fifty or seventy-five singers, led by Mr. C. E. Ashcroft as pianist. A most impressive devotional effect was produced by the union of the voices of the audience rising and singing the hymns, "Rock of ages cleft for me" and "A charge to keep I have."

On announcing the usual contribution, which has never been preceded at Music Hall by any urgent solicitations to the audience, Mr. Cook stated that the expense of opening the hall for thirteen nights had been $780. We learn that the audience has contributed about $400; that a risk of nearly $100 has been assumed by the church and society; and that three gentlemen have voluntarily taken the remaining risk, which action we have pleasure in commending as an important benefit to the public.

Having delivered, in the afternoon, before a large audience at the First Baptist Church (the use of which, with a generosity which cannot be too highly appreciated, has been tendered to the First Church since the burning of the house of worship of the latter on Christmas night), an *Address to Young Converts and to those who have united with the Church during the Year*, Mr. Cook gave at Music Hall, Sabbath evening, an address which had been announced, this being the last Sabbath for which he had made engagement to labor in this city, and his last before beginning preparations to go to Europe, as a *Farewell Discourse in Lynn*, embracing a review of the winter, with final remarks on the factories, and last words to his church and the city.

Mr. Cook occupied an hour of the evening with the topic, "Christ, our Prophet, Priest, and King; or, Christ not a Saviour if worshipped in only a Fragment of his Character," using as a text the words, "Abide in me and I in you." We greatly regret that the pressure of Mr. Cook's engagements prevents our obtaining from him the heads of this discourse, which was delivered without notes, but which would have injustice done to its argument if not reported in full. It consisted largely of a presentation of *the proof from science of the Divine Omnipresence, and of the combination of this proof with the scriptural doctrines of the Holy Spirit, the new birth, and the atonement.* A combination of the severest truths of Scripture with the religious truths to be learned from the latest science, has characterized Mr. Cook's presentation of evangelical topics in Music Hall.

We append a passage of much beauty, which was simply a subsidiary illustration in the argument, and which the speaker chanced to read from manuscript. It describes the ocean sunrise as seen from High Rock in this city.

"It was five o'clock in the late dawn, and the morning star yet hung, a sun all soul, in the eastern sky, itself soul and not sky, when I was on the lonely street on my way to the summit. Birds sang from the dewy shrubs in those short, deep warbles which it appears man is not fit to hear, but only the holy grey dawn. I was at the summit of the High Rock twenty minutes before the sun rose, and the infinite ocean lay as quietly as if its sleep were the hush following its awe after looking upward upon the innumerable worlds it had just ceased to mirror. I was watching the majestic movement of a sea bird flying toward the east, when there flashed at the edge of the sea a sparkle of flame, as if that far city, to which, in the legends, Arthur was mysteriously carried across the waves, half out of sight beyond the edge of the world, and half in sight as a glory revealing itself from below the furthest ocean's verge, had suddenly begun to burst up in conflagration. The bursting, like that of a conflagration, unequal at different points and unequal at different moments, grew, until the whole unseen infinite city seemed lifting itself in flames; and lifted itself more and more, until suddenly the conflagration that had seemed to cling to the ocean's rim, began to break loose from it, yet withdrew only little by little, but at last hung free in the infinite ether, and the flames became a sun. The lines of delicate mist that hung along the coasts took irradiation. The cities were not awake, but the chorus of birds grew louder. The poorest cottages glowed in the light not less than the loftiest mansions. The sunlight once more hung upon the just and the unjust, highest blessing or highest torture either now or ultimately, according as life is true or false; but giving every soul one more chance to make life true. The morning star withdrew its face."

Mr. Cook then occupied an hour in considering the new duties thrown upon New England by the extraordinary growth of the manufacturing districts, and in a searching, courteous, and impartial review of the winter, and in last words to his church and to the city. We shall print next week, or in our next issue, these remarks in full.

We subjoin a list of the topics which have now been presented at Music Hall. It will be seen that the variety of the topics equals their importance. The subsidiary remarks on the factories occurred at the opening of the third and fourth, and at the close of the eighth and thirteenth meetings.

1. "Similarity of Feeling with God, a Condition of Salvation." Jan. 15.

2. "The Moral Perils of the Present Factory System of Lynn." Jan. 22.

3. False and True, Faith and Repentance." Jan. 29.

4. "Christ's Gift of the Fulness of Joy; or, The First Principles of a Scientific Christianity." Feb. 5.

5. "Treason toward God; or, Christ's Human Nature the Ideal of Excellence for Young Men." Feb. 12.

6. "Secret Sins as an Obstacle to Conversion." Feb. 19.

7. "Poor Clothes, Rear Pews, and Hard Work, as Excuses for not Attending Church; or, The Value to the People of Sunday Kept Holy." March 5.

8. "The Necessity of the Atonement, an Inference from the Nature of Conscience and from the Unchangeableness of the Past." March 12.

9. "Christ in the Family and in all the Relations of the Social to the Religious Life; or, Cheap Homes and Boarding-Houses in Lynn." March 19.

10. "Steps toward the New Birth; or, Recent Science on the Signs of the Vices, and on the Best Methods of Escape from Bad Habits." March 26.

11. "Swindling Public Amusements and the Street School; or, Christianity and Science at one as to the Temptations of Cities." April 2.

12. "Club-Rooms in Lynn, Gambling-Rooms, and Dwelling-house Sabbath Drinking Parties; or, The Duty of making Christian Public Sentiment in Cities a Police behind the Police." April 16.

13. "Christ our Prophet, Priest, and King; or, Christ not a Saviour if Worshipped in only a Fragment of His Character." April 30.

The public will please understand that we are responsible for the reports of Mr. Cook's Music Hall Lectures which have appeared in the *Reporter*. Mr. Cook has furnished the manuscript of the heads of his remarks and nothing more; and these outlines used by us, and an account of the meetings written by us, have appeared at our own option, and not by his request.—*Semi-Weekly Reporter*, May 3.

CHANGES IN NEW ENGLAND IN THE LAST FIFTY YEARS.

When Edmund Burke was a young man he wrote a letter to a friend, stating that he had a plan of going to America for life; and this because the Western Continent was sure to be the seat of a great nation; was in the infancy of great changes; and was, therefore, a field in which effort put forth early would have usefulness on a great scale. New England is to-day in the infancy of great changes. It may be that if Edmund Burke were alive he would think the new opportunity of usefulness not inferior to the old.

1. The population of the manufacturing districts of New England is increasing with extraordinary rapidity.

It is the recommendation of Demosthenes that all speeches should begin with an incontrovertible proposition. It is incontrovertible that this city, as one manufacturing centre in New England, has passed through great changes in the last twenty years. I should have exhibited but a callous sensitiveness to the grave responsibilities of public speech, if I could have forgotten for an instant, in discussing the future of this city, that in the last twenty years your population has doubled, the value of the products of your chief industry trebled, and the amount of your taxable property quadrupled. It is fabled by the poets that whoever will put his ear to the ground in the thrifty and jubilant days of that season of the year we are now entering, may hear the noise of growing things. Lynn is a meadow slope on which the sun of secular prosperity has shone so brightly for the last twenty years that there is hardly a sod of it on which you can put your ear without hearing the sound of growing things.

But, is this growth a merely local phenomenon? The last census has an astonishing answer to make. The growth is characteristic of nearly every manufacturing centre in Eastern Massachusetts; and I had almost said of every such centre from Long Island Sound to the White Mountains, and from Cape Cod to the Berkshire Hills.

Take the seven cities on the Merrimac River. I often hang in imagination above that stream as the best emblem of the life of Eastern New England. Child of the White Mountains and of the Pemigewasset, the Merrimac rushes past the innumerable spindles of seven cities to the sea — Concord, Manchester, Nashua, Lowell, Lawrence, Haverhill, and Newburyport — doing more work than any other river of its size in the world; and emblematic from source to mouth of the industrial life of the Atlantic slope of New England, which more and more rushes beneath factory wheels through all its vexed course from its source in the mountains to its home in the ocean. These seven cities, in the last twenty years, have, in the aggregate, more than doubled in wealth and population.

Lawrence has grown from a pasture to a princely manufacturing centre within twenty years.

Draw a line north and south cutting the population of Massachusetts in halves and through what point does it now pass? Draw another east and west, cutting the population in halves, and where does that line fall? The intersection of the two lines is the centre of population. The centre of territory is within the limits of the city of Worcester, on the easterly side, near Lake Quinsagamond. But where is the centre of population? Is it Framingham? Is it Lake Cochituate? The north and south line which cuts the population of Massachusetts in halves passes easterly of a point midway between Harvard University and the west end of West Boston Bridge. The east and west line dividing the population into equal portions passes through the South Boston end of the Federal Street Bridge. The two lines intersect at a point not two miles west of the State House. This, according to the State documents, was the centre of population in 1865.[1] The centralization of wealth is even more remarkable than that of the population. The census everywhere reveals the fact that, through the aid of the wonderful increase of all means of intercommunication, the change which is constantly giving greater and greater power to cities, this added weight of the Atlantic slope of the State is chiefly an effect of the extraordinary growth of the manufacturing centres of Eastern Massachusetts. Of these, Boston itself is one. I must be pardoned for considering it a suggestive circumstance that, in spite of the remarkable advances of Central and Western Massachusetts, the circumscribing line drawn from the State House, and containing half the population of the Commonwealth, has contracted its radius ten miles in fifty years. All Eastern Massachusetts is a factory. In 1865, more than one half of the population of Massachusetts, seven tenths of the personal property, and two thirds of the real estate, were situated within twenty-five miles of the State House at Boston![2] In the five years since these

[1] Abstract of the Census of 1865, with Remarks on the same and Supplementary Tables, Prepared under the direction of Oliver Warner, Secretary of the Commonwealth, p. 274.

[2] Ibid., p. 275.

astonishing estimates were made, your city has increased thirty-six per cent in population. But Lawrence has increased thirty-two in the same time; Lowell has increased thirty-one; Haverhill nineteen, and Fall River forty.

Here is the incoming of an Atlantic tide. It is the roar of the industrial conditions of Old England coming into New England. I have lived for a year within hearing of the roar of the ocean. I have looked daily upon the coming in of the vast tides. It is little to say that I profess to have lived also within hearing of the roar of the human ocean which beats on the Atlantic slope of New England; and have looked daily upon the coming in of these vast tides. Imagine the magnificent coast-line from Newfoundland to New York beaten in all its coves and headlands by incoming Atlantic waves. A feeble occupation this, compared with imagining the same coast, beaten, as it is, in all its coves and headlands, and likely to be beaten more and more furiously as the years pass, by these incoming human tides, and more and more complicated industrial conditions. Not discuss those conditions! Not secure the best life that can be secured for the millions whose future is now being largely determined by the precedents which are to be set in the period of transition New England is passing! Not turn public discussion and legislation early to the solution of problems more vital than any others in the secular life of New England, and sure to become more and more complicated as the tides rise higher! He who says this is likely to be as little regarded as the rattling of rushes before the coming in of an Atlantic surge. (Applause).

2. It is notorious that, while the population of the manufacturing centres of New England is increasing with extraordinary rapidity, that of the agricultural and commercial districts is fluctuating, and, in many cases, on the decrease.

3. The distinctions between rich and poor are becoming wider in the manufacturing districts.

(1) This is partly the unavoidable result of the growth of the power of capital. (2) It is in part the consequence of the massing of men in cities, as distinct from small towns.

(3) It is in part the natural effect of the organization of manufacturing industry in great corporations on the one hand, and an operative population on the other. (4.) It is also in large measure the result of the fact that in the manufacturing districts of New England, a vastly greater proportion of the population is now of foreign descent than fifty years ago.

For nearly a century and a half the people of New England, consisting in 1640 of only about 4,000 families or 20,000 persons, multiplied on their own soil in remarkable seclusion from other communities. Bancroft points out that after the first fifteen years following the landing of the Pilgrims, there was never any considerable increase from England. Palfrey makes prominent the circumstance that it is chiefly within the last forty years that the foreigners have come. It is not true to say that New England is becoming New Ireland. But it is hardly epigrammatic to say that manufacturing New England is becoming New Ireland. Out of every hundred of the population, the number of foreign born was in 1865 in Lowell 30; in Fall River, 31; in Boston, 34; in Lawrence, 42.[1]

4. As soon as our population is thick enough, large parts of the industrial difficulties of Old England are likely, in spite of American Institutions, to arise in New England.

It is not safe to trust the action of democratic institutions to solve the problems which a monarchy as free as England has been unable to solve. What is the ministry in Great Britain but a committee of Parliament, obliged to lay down its power whenever it cannot command the support of a majority? There is no such difference between American and English political institutions, as will prevent the larger part of the questions that trouble the industrial relations of England from crossing the Atlantic. We have, indeed, no aristocracy based on blood. But it is notorious that, in spite of every feudal inheritance in English social and political life, the aristocracy based on land and manufactures and other forms of

[1] Census of Massachusetts for 1865, Abstract of, p. 298.

wealth has more power in England to-day than that based on hereditary descent.

5. The sooner New England begins to prepare remedies for its industrial difficulties the better.

(1) The legislation of England for a hundred years has demonstrated the necessity of rectifying abuses in manufacturing centres by investigations conducted under the authority of law. (2) The best literature of England for the last hundred years turns on no points more vital than the condition of the poor and the relations of employers and employed. (3) I maintain that it is greatly to the honor of Massachusetts to have organized, first of all the States, a Bureau of Labor.

These five are plain, large, and incontrovertible propositions. I confess that I am moved by them to think that, while all old duties remain as sacred as ever, and while some old duties are becoming more sacred than ever,

"New occasions teach new duties, we ourselves must Pilgrims be,
Launch our Mayflowers, and steer boldly through the desperate winter sea."

The changes in New England in the last fifty years! They expose by more than glimpses new duties thrown upon the churches, the press, the platform, the parlors and the workrooms of New England. A very dull and unpoetic fact, it may seem, that New England is inevitably to be a factory. It is one already. Plymouth Rock began the first New England. Shall it be the corner-stone of the second? Goethe listened to the spindles of Manchester and said, with profound meaning, that he thought theirs the most suggestively poetic sound of the century. He foresaw the future industrial conditions of the cities of the world, and yearned for the unborn ten thousand times ten thousand for whom that sound was a prophetic note. But all men have not Goethe's ears. When you and I are no longer in the world, the problem of the future of New England will be how to make Plymouth Rock the corner-stone of a factory. Who are the masons that shall lay that jagged and fundamental Rock smooth with the other stones of the wall? And if the lines of the Rock and of the other stones of the wall should not be the same, shall the other stones be hewn, or Plymouth Rock be hewn, to bring out the line? God send us no future into which Plymouth Rock cannot be built unhewn! But how to build that Rock into the complicated lines of the industrial relations of the future of the cities with which New England is to be crowded? How to make Plymouth Rock the corner-stone of a factory? I admit that the question, except to Goethe's ears, has almost the sound of sacrilege. But the Sphinx of current history, daughter of Typhon and Chimaera as was the Sphinx of old, already asks it in whispers; will ask it in the next generation in bold speech; and in the next century with the voice of many waters; and the penalties of not answering are not likely to be inconsiderable!

REVIEW OF THE WINTER.

Some weeks ago I said I felt like Bismark before Paris. It was a facetious remark. But I may now make one of more significance. I feel like Bismark in Paris. (Applause). Three important results have been accomplished:

1. A subtle and large evil has been exposed;
2. Remedies for it have been suggested;
3. Public sentiment has been carried overwhelmingly in favor of those remedies.

For any who wish an argument for the two measures of Factory Reform which have been recommended, I offer the fact that in this city, where the subject is better understood than it can be elsewhere, local public sentiment has overwhelmingly favored those measures.

Thucydides and Talleyrand, though as widely separated in character as in time, both advise that those who would deal with the public usefully or successfully should practice entire freedom from indirection. The secret of securing attention, as well as of speaking usefully, is to say the thing that needs to be said.

There are several facts of public notoriety here which are now matters of the past. I judge that the past is not likely to be reversed.

1. It has now been my fortune to speak in Music Hall for more than a quarter of a year. It is only two weeks less than a third of a year since the first of the Music Hall Services.

2. In January, I made here certain very damaging charges concerning the condition of your chief industry.

3. The charges were such that beyond doubt I should have been advised, if I had consulted beforehand with any one not wise as a serpent but harmless as a dove, not to present them.

4. So damaging were the charges and so public and unequivocal was their presentation, that, if they had been untrue, or if they had contained substantial exaggeration, I should have been made the most unpopular man in public life in this city, and months ago have been hissed off this platform.

5. I have not been hissed off the platform. I am saying nothing of the necessity of repressing applause here night after night for a quarter of a year; nor of a deliberate request from the class said to have been slandered, for the continuance of the meetings; nor of scores of other indications. In a gathering of some sixty-five operatives in this city last winter, not long after the topic of the factories had been brought to public attention, a vote was taken on the question of the advisability of the separation of the sexes. Two speeches were made against the measure; and the vote stood for it, sixty-three to two. The assertion that the proportion of these figures to each other fairly represents the present sentiment of the fifteen thousand working people here, I make no use of, except to say that it comes to me from an exceedingly large number of the most respectable sources, and that no counterbalancing evidence reaches me. It is not consonant with the rules of local Crispin lodges to act on certain themes without a kind of concurrent action or permission from national or international lodges; otherwise, resolutions which have been twice prepared by working men to be brought before the local lodges here on this theme, would have been passed. I notice that the admirable Report of the

Massachusetts Bureau of Labor, published but a few days since for the use of the Legislature, has a passage comparing the moral perils of the old and new system of your chief industry; and bases an inference that the perils of the latter are the greater, on the fluctuations of the business, the very point urged three months since in this Hall. These and many other similar facts it is not necessary to bring into employment. It is enough to say simply that I have not been hissed off the platform. Every man of sense who heard what I said in January knows that I should have been, if I had not spoken substantially without exaggeration. As to the applause, petition, and other indications of local sentiment, they mean as much as they would on the other side. How much is that? As much as they would on the other side, if that side had ever received them. (Applause.) When poor John Sterling, Thomas Carlyle's friend, after a life tossed with scepticism, lay dying, he said: "Bring me the Bible I used to use in visits among the cottages at Herstmonceaux," and laid his head on it, as if in rest from all his tossing. I will remember these two hundred signatures of working men, as John Sterling remembered the Bible he used among the cottages at Herstmonceaux.

6. Nearly four months have now passed. There has been abundant time for sentiment to become calm and intelligent. Photograph public sentiment as it stood two months ago. It was overwhelmingly on the right side; but it would have been unfair to take the picture to fix the record then. The picture was too near and would have been blurred. Take the photograph to fix the record two years hence, when all the facts will not be fresh in memory, and the picture would be too dim. Now is the time to take the photograph to fix the record. There has been no especial excitement in the city for the last month.

The photograph is fixed by these six propositions. I recite them not for my own sake, but as an incontrovertible indication of local public sentiment. The propositions are all matters utterly undeniable. They are such as a stranger looking on from outside, might be sure of. But whoever will run his eye along the six, and take their trend, will see that an inference follows from them as to the nature of local public sentiment. It is that local public sentiment which I to-night add to all the arguments I have heretofore advanced.

And now it is my duty to thank this city, which is dear to me, and will continue to be dear, as are the ruddy drops that visit a glad heart. You have not been misled. You have not been intimidated. There are many ways in which Capital may act as a Ku Klux Klan. Capitalists range all the way from the George Peabodys to the James Fisks. Years ago there came into classic use in this Commonwealth the significant word gerrymandered. To divide a state into political districts in such a manner as to favor a particular political party, was a scheme which, under Elbridge Gerry's administration, was forever put into the pillory by that single word. I have met, in one of the most scholarly publications of Boston, the word gymfiscation. It has as good a right to become a part of the language as the word gerrymandered. You are yet a young city. You have yet but a comparatively small population. I say and beg you to notice that I say that I think you have no one or two, or any other number, as corporations or as individuals, near, or nigh to near, to that rank of capitalists which the James Fisks represent at the end of the scale opposite to the George Peabodys at the other. But you will have them in the future. Let me beseech you to remember that all your difficulties will assuredly increase with a more crowded population. I am not thinking of your past. I am thinking only of your future. Charles Dickens regarded it as a peculiar vice of American cities that they are managed almost exclusively by local capitalists. In his celebrated American Notes he pointed this out as the peculiar shame of Americans. As I hope the city will never be politically gerrymandered, so I hope New Lynn will never be gymfiscated. I thank the singers. I thank the newspapers. I thank the public sentiment of the churches.

And now, to my own church, God send the spotless robes and the queenly spirit of a Bride of Christ, and enable her to carry them untarnished through all the dusty ways of the future of this city. My whole motive for a year, and a more solemn one could not lie upon man, has been to do my whole duty as a minister of a church exposed to the fierce temptations of a crowded population in a manufacturing centre. I was resolved that the church should be put on its knees neither to Labor on the one hand, nor to Capital on the other, nor to City Vices in any form; but on its knees only to Christ. The topics selected by me in this Hall have, for evident reasons, been off the range of my usual courses of public remark. It has been my fortune to pass the last three years almost exclusively in speaking from point to point of New England in revivals. In the first half of the year which I have passed in this city, there was a revival in the church. This was in the heat of summer; and it was my plan to fill the whole winter with the most severe labor in a revival. Some fourteen had united with the church by profession of faith. But on Christmas night the church edifice was burned. That event of necessity changed the plan of the winter. Although all the churches of the city held out their arms full of generosity to us; and although one of them, with a generosity which it will take long to repay, has held us in its arms every Sabbath since our calamity, it was found, after careful inquiry, that there was no place open to the church where Sabbath evening devotional meetings could be conveniently held. The devotional meetings had exhibited, up to the last, the best state; had more than trebled in size; and it was my intention that you should have done severe work in them for the whole winter. We were driven to this Hall. When the fire took place and we came here, the church was lifted for half of each Sabbath from the position of a wax taper in one parlor at West Lynn to that of a watch-tower for the whole city. I thank the church for the resolution it passed unanimously after my address of January 22, when public opposition was offered to the performance of its plain duty. You understood perfectly, when you passed your resolution, that a pulpit which is allowed by the parish that owns it to be plastered over with directions from outside the parish, or whittled by knives not sharpened on the foundation stone Jesus Christ, becomes such a symbol that it may be difficult to find a man to stand behind the whittled thing. I was told in Boston the other day by a lawyer, that matters had been carried so far that two and perhaps more parties here owed their freedom from the clutches of the law exclusively to my forbearance. I have never replied to attack on myself. When your interests have been involved, when public interests of the city have been involved, when the freedom of public discussion has been involved, I have made reply. God send you in your new church edifice a revival of that deep and calm kind which was going forward in the old when the flames came; and may the increase of that religious thoughtfulness and enthusiasm fill the new temple as the flames filled the old. God send you a minister whom no one can vex, or cajole, or browbeat, or outwit. I love this church; and, when I lie dying, I will remember that we labored together for the poor. And now my face is toward Europe. When from the deck of the steamer I look toward Lynn from Massachusetts Bay, as the gates of the ocean draw near, the multitudinous waves shall not be more abounding than the good I will wish you and this city, until we meet, when the heavens are no more. The supreme duty is to follow the Pillar of Fire.

A letter has been placed in my hands which I will read and answer here in the serious presence of this assembly.

LYNN, April 29, 1871.

REV. JOSEPH COOK:

Dear Friend and Pastor: Please accept this picture and these books from a few of your friends in the First Congregational Church and Society in Lynn, who, painfully conscious of the inadequacy of words, or of any token to express the sentiments of high regard which they entertain for you personally, and for the labors you have performed in the endeavor to promote their eternal welfare and that of others, present them to you, that in the future they may remind you of those who desire not to be forgotten yet dare not presume that, in the eventful and eminently useful life which they anticipate for you, they shall preserve a place in your memory without something constantly near you to represent each of them, and to say for each, "Remember me."

Mr. and Mrs. J. F. PATTEN,
Mr. and Mrs. G. H. CHADWELL,
Mr. and Mrs. CHAS. E. ASHCROFT,
Mr. GEORGE F. HOSMER,
Committee.

On returning yesterday afternoon from the crowded delight of finishing a tour of two hundred calls in the church and society, I was met at my rooms by the entire surprise of your costly and most beautiful gifts: a magnificent picture, called "The Sealers Crushed by the Icebergs," and valued at ninety dollars; the volumes entitled "The Circle of the Sciences," valued at twenty-five dollars; and "The English Hexapla," in the most valuable of the English editions, valued at twenty dollars. I have received private assurances that the plan of making these gifts was of the most spontaneous kind both in conception and in execution.

I can bear opposition; but it is not so easy to bear kindness. I confess that I felt unmanned by this blow; and do not think that I shall soon recover from it!

A city church is either the best or the worst place to grow Christians in. If, in a crowded population, where the temptations of a church are the fiercest, these are resisted, the church becomes strong in proportion to the fierceness of its temptations. But, if they are not resisted, it is a wholly familiar teaching of Church History as well as of the Scriptures, that a church may become one of the most withered, juiceless and jejune of all the beads of growth on God's vine. Your gifts, occuring at the end of a year passed one third in a revival and one third in illustrations of the spirit of that revival in public discussions of topics of importance to New Lynn and New England, are invaluable as indications that you purpose that your church shall resist all the temptations incident to a crowded population. You mean that in no sense shall it be put on its knees to Labor on the one hand, or to Capital on the other, or to City Vices in any form; but on its knees only to Christ. "Scratch a Russian," said Napoleon, "and you will find beneath the surface a Tartar." Scratch an inefficient city church, and you will find beneath the surface a club. You mean that yours shall be a church, and not a club. You mean that it shall be as careful of Orthodoxy in Practice as of Orthodoxy in Doctrine. You mean that it shall not be unmindful of the teaching and the trials it has had in a long and not undistinguished past. Certain most serious and persistent misrepresentations of the history of the church for the last four months will be set at rest by these gifts. They are priceless to me for their own sakes, for the sake of these principles, and for your sakes. You have given me a picture of which the name is, The Sealers Crushed by the Icebergs. The true name of the picture of the winter, as you will allow me to say reverently, for the sake of New Lynn and New England, is, The Icebergs Crushed by the Sealers. (Applause.)

Mr. Cook's Farewell at Music Hall.

The thirteenth of the series of Sunday evening services which have been held at Music Hall by Rev. Mr. Cook was attended with unusual interest on Sunday last; it being the occasion of his last meeting in Lynn before beginning preparations for his European tour. A very large congregation was present. Vases of choice flowers were tastefully arranged upon the platform, and the chorus of singers was numerically increased. The opening exercises were impressive. Before the taking of the usual contribution, Mr. Cook thought there were some present who would deem it a favor to know the financial status of the Music Hall meetings. The expense of the hall for the thirteen meetings had been about seven hundred and eighty dollars; the audience had contributed nearly four hundred dollars; and three gentlemen had volunteered to become responsible for whatever arrears there might be at the close of the meetings......

We did not learn the amount of the last contribution.

Mr. Cook chose as a text, John xv. 4, "Abide in me, and I in you," and announced his subject as, "Christ our Prophet, Priest, and King; or, Christ not a Saviour if worshipped in only a Fragment of his Nature." To abide in him, we must not only accept him as Redeemer, but follow his precepts. We should be orthodox in practice, as well as in theory. We could not rely alone for salvation on the doctrines of election, the universality of the atonement, the unlimited grace of God, etc., for that would not be worshipping him in the whole of his divine nature. To abide in him, we should know what is meant by a love of God's works. He reveals himself as a power behind all the works of nature. He is as plainly visible in the humblest flower that grows beneath our feet as in more pretentious designs. We should realize that wherever God acts, there he is. Mr. Cook enforced the idea of Divine Omnipresence by reference to the works of nature in general, and dwelt particularly on "Neptune's Cup," a coral in the Indian Ocean formed symmetrically in the shape of a goblet, indicating the presence of an intelligence behind and beyond the simple instinct of the insects whose united existences, although wholly independent of each other, finally resulted in this beautiful design.

At the conclusion of the sermon, which was both able and lengthy, Mr. Cook devoted an hour to Final Remarks on the Factories, — the duties arising from the rapid growth of the manufacturing centres of New England, — embracing a general review of the winter, and last words to his church and the city. Whatever else might hereafter be said, it could never be denied that thirteen meetings had been held in Music Hall; the past cannot be changed. And the general applause which the sentiments advanced at those meetings had received amounted to *as much* in their favor as such applause for the opposite sentiments would have done for *them*, had the opposite side ever received it. On a former occasion he had said that he felt like Bismarck before Paris. It was a facetious remark. He wished to-night to make one of more import: He felt like Bismarck *in* Paris. To have taken a photograph of Lynn, for the reflection of her public sentiment in regard to the questions which had been discussed in Music Hall, at an earlier date, would not have been fair — the picture would have been blurred by the nearness of the events. To wait a few years, and then take the photograph for that purpose, would be equally unjust — the picture would be too dim, from the remoteness of the events. The proper time to photograph Lynn was now, after nearly a third of a year had passed since the beginning of these meetings, and when the feelings and reason are calm and dispassionate. So far from having been supported by this public sentiment, had there been no truth in the charges which he made there in January, he would have been hissed from the platform.

Upon returning home one day last week, he was surprised to find several costly presents, accompanied with a letter, from his friends, as tokens of their esteem for him, and as mementoes of their pleasant connections in Lynn. These gifts consisted of a splendid chromo, from the celebrated painting of William Bradford, "The Sealers Crushed by the Icebergs," valued at ninety dollars; two volumes entitled "The Circle of the Sciences," valued at twenty-five dollars; and a copy of the "English Hexapla," embracing the Greek New Testament, with the six most important English translations, valued at twenty dollars. Mr. Cook read the letter accompanying the gifts, and closed his remarks by reading his in reply. The two letters denoted the existence of a strong attachment between the pastor and his people. —*Transcript*, May 6.

VALUABLE TESTIMONIAL.—Rev. Joseph Cook has been made the recipient of a costly and beautiful testimonial of the esteem in which he is held by his many friends in this city, consisting of a magnificent chromo from Bradford's celebrated painting of "The Sealers Crushed by the Icebergs," valued at ninety dollars, and three books, comprising two volumes entitled "The Circle of the Sciences," valued at twenty-five dollars; and a copy of the "English Hexapla," embracing the Greek New Testament, with the six most important English translations, being the most valuable of the English editions, and valued at twenty dollars. The presents are on exhibition at the Jewelry store of Mr. John Carruthers, on Market street. The gift was of an entirely voluntary character, and the contributions made without solicitation after the project became known. Mr. Cook read, at the close of the Music Hall service, a letter in reply to the gifts; and this will appear with our full report of his remarks in the latter portion of the meeting. — *Semi-Weekly Reporter*, May 3.

MR. EDITOR: — I learn that Flavius Josephus Cook is about to close his career in Lynn. I have waited to the utmost limit of time for an apology from him in relation to the unjust attack in which he slanderously maligned and injured those unprotected and innocent women who work in a factory which belongs to me — hoping that common decency would prompt him to repair an injury that he had done, so far as was in his power. He doggedly and persistently refuses to do so. The language he uses relative to them is so vile and obscene that I cannot repeat it, and could emanate only from a heart wicked and malicious. And if he leaves town with this hanging over him — when they have kindly offered to forgive him, if he will only make a square apology — it should follow him wherever he goes.

I learn that he talks of leaving our shores for Germany, because, having graduated from several seminaries here, and got all the learning that our country affords, he can learn nothing more here. Now, Flavius Josephus, I think there are many things more you can learn in this country. I think you could learn to be modest, and possibly might learn to be gentlemanly, should you make it your study. Learn to be pure in thought and word; learn to be agreeable in manner; learn to honor and respect woman. It betrays a defect in early education to speak disrespectfully of woman, and the kindly affection one should cherish towards his mother would seem to forbid it.

You have repeatedly said that you felt like Bismark before Paris, and last that you felt like Bismark *in* Paris, exulting in victory. The only victory you have gained is over these innocent and unprotected women. To exult in that bespeaks a spirit not of heaven or of earth. The rest of your career has been one complete defeat. You have ill-treated your own people, and insulted every one who dared to differ from you. I hope you leave Lynn a wiser if not a better man.

S. M. BUBIER.

Semi-Weekly Reporter, May 6.

A QUESTION OF PRIVILEGE.

MR. EDITOR: I am well aware of the editorial rule that forbids discussion, in his own columns, of anything written by the editor; but I do not believe that any conductor of a paper who intends to be fair will insist on such a rule, if any person should feel aggrieved, or desire to make an explanation or correction of a given statement. But the point I am about to propose has a broader scope than any merely personal consideration, great and vital as such questions are liable to be. I shall claim the right to speak freely, earnestly, and as fully, within reasonable bounds, as I think the subject demands; and I trust I shall not be hampered by any conventional rule which might cramp a conscientious statement of facts or expression of sentiments.

I suppose I shall be understood, if I remark, at the outset, that "Mr. Cook," as he is usually called in the papers, has been stopping in Lynn for some months past, and has been conducting certain "services," as they are denominated, at Music Hall. And in confirmation of this, I may be permitted to remind the public that in February a man was sent to Salem Jail for trying to say a few words, one night, at one of these meetings; the Police Court of this city consigning him, without hesitation, to that retirement, on the charge that he was "disturbing public worship"; and the man having neither money (to pay a fine), friends, nor counsel to test the case, it practically went by default.

It is perfectly notorious that this man, this Mr. Cook, has caused more bitterness, more strife, more scandal, more foul thoughts, uttered or unexpressed, within the time he has been holding these meetings, than any other man, and than all other men, so far as can be computed, that have ever lived in Lynn. Furthermore, and it may be a satisfaction to him to know it, he has caused Lynn more injury to its good name, abroad, especially where it is least known,—though it has dealings with every city and considerable town in America,—than any man who ever partook of and abused its hospitality. And this, too, in the shortest space of time. Of this injury abundant evidence can be furnished, if the statement is questioned.

I say, this language is not too strong. It ought not to be suppressed, and it cannot be gainsaid. Mr. Cook is a public man; he advertises as such; and he should not shrink from being criticised as such. With regard to the damage he has actually done to Lynn, that cannot be positively estimated. Perhaps, in the attempt, it *has* been overestimated. As far as the mercantile credit of the city is concerned, he could not touch it; but as to that which is dearer to every community, as well as to every man and woman, — a good name — he has succeeded in causing foul suspicion, as I have just said, where we are known the least, by his reiterated accusations and inuendos. It matters not, even if his motives were not the worst; it is not necessary to charge *that.* If a man injures you, and persists in doing so, against your remonstrance and that of all your friends, he may say that he wants to reform you, and adopts this eccentric method; but it is hard to convince you, or anybody else, of such a proposition.

What has been Mr. Cook's ultimate object? Nobody can deny that he has sought popularity, at least notoriety, and that at these "services" he has, under an absurd disguise, courted the applause — "restrained applause" was the kind his aesthetic temperament most craved — and even the laughter of the thoughtless.

To attempt to go into this general subject to any extent, even to the recapitulation of the titles of his "thirteen discourses," would be tedious and disgusting, and we will not differ as to their omission. Nor is it my intention to allude to the heartless and unfounded insinuations he has cast, upon the working men, and especially women, of Lynn. I do not think I exaggerate in saying that if any other man, particularly some "rich manufacturer," as Mr. Cook would say, — albeit he boasts of his own sufficient riches,[1] — had been mad enough to say the things which he has said, the mildest punishment which would probably have been meted out to him would have been to hiss him from the stage, and to follow him with malignant persecution through his business and social relations.

Mr. Editor, in concluding these reflections, I must not omit to speak directly to my title and text, even although it is (but perhaps ought not to be) a delicate one. I refer to the course of the *Reporter* in connection with Mr. Cook's Music Hall Lectures. I will give it credit for suppressing entirely the most sensational and disgusting discourse, although with the remark that it was "the design to have it put up in cheap pamphlet form for the use of those interested. It will be ready in a few days."[1] And also of skipping and cutting some of the most objectionable language and ideas in several of the others. But the fact remains, that the oldest and oftenest-printed newspaper in this city has continuously thrust before its readers two or three columns a week of matter, much of which — to state it very mildly — was offensive to good taste, good feeling, and good manners. The extenuating plea — which it is to be hoped will not be made — that "the public demanded it," is the stereotyped one of the "Satanic Press" of New York, or other large city, in defence of publishing the full details of prize fights, or of certain criminal cases. No people in Lynn have yet demanded the pamphlet just spoken of; but, on the con-

[1] In one of these lectures Mr. Cook says substantially as follows: "And had I had a timid wife pulling me back, I should still have pursued the same course I have pursued. . . . . I am not dependent upon my salary for support." As a matter of fact it is proper to remark that he never had a wife, and he appears to be, in those respects unlike Daniel Webster, who said he was "neither rich nor a bachelor."

[1] It was never done, and never will be.

trary, many have protested against its publication. There has never been any considerable demand for the omitted passages, as, probably, few people were depraved enough to wish to file them away as choice reading, or as valuable facts.

Finally, I know that I but express the feelings and views of all with whom I have spoken on the subject, in sadly regretting that the history of Lynn, for such our press is continually recording, should be cumbered with much of the matter from which it just now suffers. But what is writ, is writ; and while this Mr. Cook has filled a score or so of columns in which the attributes of charity, or even of manliness, are among the least conspicuous omissions, it should and shall be said that an earnest protest and indignant denunciation and denial accompanied the record of his ignoble task.

A Native of Lynn.

Semi-Weekly Reporter, May 17.

"A Question of Privilege."

It is, undoubtedly, a great privilege to be able to scold an editor through the columns of his own paper. Our correspondent, "A Native of Lynn," has done this, — we hope to his heart's content, — and we trust he feels better for it. He certainly ought, after having thus freely given us "a piece of his mind." Under some circumstances, we should have treated his effusion simply as a mass of impudence from an ill-bred clown; but, knowing the writer, as we do, to be a gentleman and a scholar, who, though meaning well, is just as likely to err as other men, we have admitted his "question of privilege," leaving it to our readers to say whether he has abused it or not. We have not spent thirty years of our life in the newspaper business without having learned that there are a great many people who know better how to conduct newspapers than those who do conduct them. We are glad that there is at least one such gifted individual among us. If we could only induce him to take our place, we might have leisure to go about and tell our manufacturers how to make shoes, our carpenters how to build houses, and our tailors how to make clothing. Next to this, we should like to be able to submit beforehand everything we write and everything we propose to print to the kindly criticism of a score or two of the censors of the press who exist in every intelligent community. The main objection to this would be that we should have to issue two blank sheets every week. We would like to please everybody, if we could, but long since gave that up as a hopeless task. If, by any means, such a miracle should be performed, there would still be one drawback. Our friend and such as he would have no escape-valve for their pent-up indignation, and the consequences might be disastrous. Othello's occupation would be gone entirely. But it is not likely so to be, and we must continue the even tenor of our way, striving, as we ever have, to do all the good we can in the community, and as little harm as possible. Meanwhile, we expect to be found fault with by some, and scolded by others; but shall endeavor to bear it all patiently, profiting, as far as possible, by the advice and suggestions of our friends, letting all else pass for what it is worth. It is our wish to write and print nothing that, dying, we should wish to have blotted out; but to err is human, and we boast no exemption from the ills which soul, as well as flesh, is heir to. So far as Mr. Cook is concerned, we are not his apologist or defender; he was able to take care of himself. But he has gone from among us. The evil that he has done, if any, will die with him; the good, if any, will live on. Let us have peace. — *Semi-Weekly Reporter*, May 17.

"Hearing with the Elbows."

As an evidence of the manner in which remarks made by Rev. Joseph Cook at Music Hall have been misrepresented, even by some of his auditors, we give a rather amusing instance that has come to our knowledge. In the reverend gentleman's discourse upon "The Sorceries of Secret Sin," delivered on Sunday evening, Feb. 19, he said that if the white of an egg should be placed in alcohol, it would shrivel and harden; and as the brain was composed largely of the same substance as the white of an egg — albumen — it followed that alcohol must have a similar effect upon that organ. And scientific research had proved the theory to be correct. On the following day one of his auditors gave him credit for saying that, by mixing the white of an egg with alcohol, one could manufacture human brains! There is no longer any necessity for a vacuum in one's cranium, and the fools may take courage.— *Transcript.*

Lynn, Sep. 7, 1871.

To the Rev. Joseph Cook:

Dear Sir: Hearing that you are about to publish a pamphlet upon the subjects under discussion in our city last winter, we, the undersigned, fully endorse the two great principles then considered, the separation of the sexes and the appointment of good overseers in the shops, and we believe that a very large majority of the laboring class believe that such a course would be for the best interests of all and add to the morality of our city.

[*Signatures.*]

These names are not to be made public; but, if the note is ever questioned, you are at liberty to show the signatures.

[*Signature.*]

The undersigned have been shown the above and unreservedly endorse the statements therein contained.

With unusually good opportunities of knowing the sentiments of operatives in the shoe-shops in Lynn, we are very sure that it is no overstatement of facts to say that of the fifteen thousand employed, at least ninety per cent cordially favor the measures of reform so nobly advocated in Music Hall last winter; and formal action in this regard would long ago have been taken in the various Crispin lodges, did the rules of the order not forbid.

[*Signatures.*]

Children in Factories.

We know indeed that there is a compulsory statute of the Commonwealth in relation to the schooling of its children, but, like a great many other statutes on the books, it is paralytic, effete, dead — killed by sheer neglect. It was never enforced, and never supposed to be anybody's duty to enforce it. In fact we are inclined to believe that it is not generally known that such a law was ever enacted. *Nobody looks after it, neither town authorities, nor school committees, nor local police, and the large cities and many of the towns of the State are swarming with unschooled children, vagabondizing about the streets and growing up in ignorance and to a heritage of sin. The mills all over the State, the shops in city and town, are full of children deprived of their right to such education as will fit them for the possibilities of their after-life.* Nobody thinks of either enforcement or obedience in the matter, so that between those who are ignorant of the provision, and those that "care for none of these things," thousands of the poor younglings of the State, with all her educational boasting, stand precious small chance of getting even the baldest elements of education.

And this brings up thought of *the successful experiment in England of her system of "Half-time Schools," as now established and in full success in many of her manufacturing towns.* Some account of them will be found interesting, and may prove suggestive of their value if sanctioned by law in Massachusetts. They are founded on the principle that children employed in factories must be allowed some portion of each day for educational purposes, and that their employment for a full day's work of ten hours within the walls of a mill is in excess of their physical ability, and would therefore redound to the injury of both mind and body. To remedy the evil they are allowed on each day three hours of schooling, and every factory child in the kingdom *works only half a day and attends school the other half,* and it has been abundantly proved that these "half-timers" eventually obtained as much book instruction as the children in the same schools, under the same masters and by the same methods, obtained under the full time of five and six hours. . . . .

Many an agent and superintendent are in error, when they state that children under ten years of age are not employed. We know they are; and we know that parents deceive overseers and overseers deceive superintendents and agents in this matter. A treasurer of a mill, or an agent, may issue his prohibitory order, but he cannot, without minute personal inquiry, know of its being obeyed. "He may," like Hotspur, "call spirits from the vasty deep," but he cannot be sure that they will answer the call. There are children employed there not over eight years of age, if appearances do not deceive, or unless mill-life and work eleven hours a day have dwarfed the twelve-year old down to less than ten. A puny child operative in Fall River was sent from a mill to wash a mop in a pond, and the mop, on being saturated with water, was too heavy for the child's strength and dragged it into the water, whence the little thing was taken out drowned. At the funeral her age was registered at nine years. A boy was found, belonging to another mill, old in features but small in stature, whose age we ascertained was eight years. Most of those we saw reported themselves, as "going on to eleven or twelve years." The school law is obeyed, but not the ten hour law for children between ten and fifteen years old. . . .

Something efficient must be done, and done quickly and effectively, for *ignorance in the manufacturing towns is on the rampant increase.* We venture to assert that never till within these last few years, could it be said, *that in a single establishment of about* 1,600 *working people in one town in Massachusetts, there were more than* 800 *that could neither read nor write!* And this ignorant horde is daily augmented by the imported influx of tens of thousands of thoroughly ignorant emigrants,—paupers, many of them, as declared by a manufacturer at a public hearing before a legislative committee, paupers imported from England expressly to be employed in manufacturing, because of the cheapness of their labor. Our large cities and manufacturing centres are surcharged with younglings growing up in ignorance and to a heritage of crime, notwithstanding all our appliances of education. And one strong reason among many others, is that there is no enforcement of the school laws. We boast of these laws and of our system of education and of its results, but the ratio of ignorance is increasing beyond the proportion of its means of cure, so that we are driven to the conclusion that the State, for its own sake, for its reputation's sake, a reputation that cannot much longer be maintained by self-laudation, for the sake of its whole people, must adopt, and put into systematic and rigid enforcement, measures absolutely positive, for the education of the very large number of its untaught children......

Factory System at Lynn.

In twenty-eight returns the length of the working season is given as ranging between eight and one-half and twelve months, the average being ten and one-half months. In the great centres of the trade, Lynn and Haverhill, the working season is considerably shorter, being, perhaps, fairly represented by eight and one-half or nine months. We give it as our judgment that a general average of nine and one-half months is not far out of the way, with a constant tendency toward shorter seasons as machinery improves in variety and excellence.

The manufacturers (33 in number), reporting their pay-roll for six months from January to July, give the average pay per person employed at $1.83 per day.

The earnings reported do not admit much margin of saving to the workman year by year, and the facts show that it is becoming more and more difficult to secure homes by wage accumulations in this branch of industry. In some of the country towns, a considerable number are able to live in houses of their own, some of them inherited, some of them bought with bounty money, and others partially or wholly acquired by wage earnings......

An investigation of the strike at Lynn elicited much valuable information. The factory system has there found the fullest development, with its numerous drawbacks and compensations. Work in all the departments is largely done by machinery. Each class of labor devotes itself to a specialty, and a workman taken from one department and put into another, would fail to make headway, or earn living wages. The use of machinery has virtually swept away the old race of shoemakers who could make up an entire shoe. As mentioned elsewhere, the work is confined to two seasons, and between them there is an interval of three or four months, and sometimes even a longer time. As the season opens, a man will get three or four days' work a week, with gradual increase till the rush comes; this in turn passes off, and, work beginning to slack, it at last tapers off to nothing at all. . . . . . "Last year," says a workman, "I had all I wanted to do from January 1 to May 1. From May 1 until the last week in July, leaving out the first week of the former month and the last of the latter month, eleven weeks in all, I earned a few cents over $36."

"Since the old system of working in little shops was abandoned," says another, "for that of large manufactories, there has been a steady diminution in the length of the working season per year. Before the time of factories there would be a steady run of employment for from seven to ten years, only interrupted by commercial depressions or revulsions. The working hours would be from twelve to fifteen. The season for lighting up, was from September 20 to May 20. Since that time, there has never been a year of steady work. At first a month only would be lost; now it has got so that we lose over four months' time every year. The system is worse here than elsewhere, because machinery has been more thoroughly introduced."

In the busy season men will come into the place from their homes at a distance on the farm. They are not as skilful, reliable, or intelligent as the permanent men; but the system of subdivision admits of their employment. . . . . . There is comparatively nothing to learn, and so no apprenticeship is required.

Some think the factory system has proved injurious, as hindering the development and growth of intelligence. In the little shops one would read while the rest worked, the reading being accompanied or supplemented by conversation upon the topics treated. In the factory there is no chance to read, and the noise and hum of machinery prevent general conversation, even when the rules and discipline do not positively forbid it. By others it might be claimed that the extra time gained in the shorter working day, gave increased opportunities for special culture. It is doubtful, however, if the latter is a fair offset to the former in its educational influence, since the one system would give unconscious improvement, while the latter would require greater self-control and more persistent application.

In its effect upon morals some think the old system preferable to the new; others claim that there is no special difference between the two. Doubtless the weight of evidence is in favor of the former view, on the general ground that fluctuations in employment produce effects analogous to similar fluctuations in business, and that in a variety of ways.

The day and week hands work from 7 o'clock A. M. until 6 P. M., with an hour at noon. Others do all they have strength to do, some working from 4 A. M. until 10 or 11 P. M. The temptation is very great for a man who has been without anything to do, week after week, to work thirteen hours a day, with corresponding high wages, even if by so doing he is reduced in a few weeks to a mere skeleton. Time even to eat is not taken, under such circumstances; a hasty bite of cold lunch, or a hurried resort to a restaurant sufficing, the whole consuming, perhaps, not more than fifteen minutes of time.

Many women work at different branches of the business, but very few of them are married. The sewing machines are largely run by young women, but the work requires the best of health and strength. Boys of eight years old and upwards, are employed a good deal. They are taken in to do what men will not stop to do, when there is a rush. . . . . .

Most of the workmen live in the outskirts of the city, or in neighboring towns and villages. The rent of a comfortable house for a family of four, comes to $200 to $225 a year. Nearly all with families hire their tenements; a few have places wholly or partly paid for. One says:

"Nine out of every ten of our shoemakers who own houses, earned them under the old system, and at a cost of not more than one-third of the present value of the same or similar places. Don't know one who has paid for such a home, by the labor of his hands alone, since the factory system came into vogue. Some have obtained them through a start given by bounty money, a favorable marriage, or some other streak of good fortune. Some have made a beginning through the aid of money loaned on mortgaged security."

Another says:

"The shop's crew that dresses the best, and in all respects makes the neatest appearance of any in Lynn, is one that can be starved out the quickest, because not a man among them owns, clear of incumbrance, a house of his own."

According to wages received the workmen may be classified as well paid, moderately paid, and low paid. In the former class the earnings of a year will range between $800 and $1,200, the latter sum being very rarely earned, and never unless partially from the labor of others. The channellers, the men who run the McKay and the beading machines, and the men who finish the bottoms, have jobs by means of which they can make their expenses when others cannot. In one shop a man is paid $1,100 for a year's time, during which he will probably work nine months and have three to himself. Another for the same work gets $1,000. He is not quite so skilful as his comrade. In other branches there are men who do as well, men who run the heeling machines, for instance. All machine operators have more steady employment than men who work by hand. The lasters, trimmers, edgemakers, and heelers have less fortunate and remunerative jobs; yet an edgemaker will sometimes earn $50 in a week. And yet those that earn $45 and $50 a week at the highest, will not have received any more the year through than a hod carrier with steady employment. One man earned as a laster $1,045 in the working year, but he was one of a few, who could do the work of three ordinary men in the same time. The earnings of the medium class range from $600 to $800 per year, the average being much nearer the former than the latter figures. There is quite a class of men who work by the week. They get from $12 to $15, and are unemployed as much as the rest. The lowest paid workers earn from $200 to $600 per annum. The average, taking the mass together, would not probably be far from $600, which sum would only give a small family a hand-to-mouth subsistence.

Very few family men do more than meet their household expenses with their year's earnings, though the majority are comparatively free from debt. With twelve months of work at current prices, it would be quite possible for a man with a moderate family to lay up $200 per year; under the present circumstances he can barely come out square......

The general intelligence of the Lynn shoemakers is good. Most have a fair common school education. It is a very rare thing to find one that is compelled to make his mark on the pay roll, and the man who has the most extensive acquaintance in the Crispin organization, some two thousand strong, says he knows but one member whose name has to be written for him......

As to church-going and Sunday observance, one expresses himself to this effect;

"The shoemakers of Lynn, as a rule, do not attend church. In the summer season great numbers of them go to Nahant. Picnic teams start at nine o'clock, and run all the forenoon and afternoon. They are generally crowded every Sunday. The women and children go with them. I think, however, that an absolute majority of them spend their Sundays at home in their own houses.

"During the intervals between the seasons of work, though very many are idle, the city is as quiet and orderly as any city in the commonwealth. You will often find the men about the post-office or periodical stores, off yachting or away visiting their friends, or at home reading or studying about public affairs, perhaps receiving visits, or doing jobs about the house. There is very little actual dissipation.

"Intemperance often has its beginning in the exhaustion following overwork in the season of drive. After the season opens, and when they begin to look thin, worn and 'played out,' the men will be seen going into beer-shops for a glass of ale. Still, seventy or eighty per cent are temperate men.

"There is much enforced idleness in the busy season, in order to recruit one's temporarily exhausted energies. There are always some men out, and when you lie by, you can turn your work over to another.

"Several of the manufacturers have sprung from the bench, but most of their money has been made since the war broke out. One made $1,800 a day for awhile, when prices went up rapidly. The feeling of manufacturers towards their workmen differs; some think of them as of the men that dig their ditches or take care of their horses; others show feelings of respect, and treat them as men rather than tools." — *Extracts from Massachusetts Senate Document No.* 150, *the Second Annual Report of the Bureau of Statistics of Labor of Massachusetts, March* 1871, pp. 493, 466, 490, 239–248.

www.ingramcontent.com/pod-product-compliance
Lightning Source LLC
LaVergne TN
LVHW021403110826
845150LV00007B/1768